Yuri K. Santos
Fábio Merçon
Alcides W. S. Guarino

# Analysis of toxic metals in fish

Yuri K. Santos
Fábio Merçon
Alcides W. S. Guarino

# Analysis of toxic metals in fish

Rio de Janeiro/Brazil

ScienciaScripts

**Imprint**



Cover image: www.ingimage.com

This book is a translation from the original published under ISBN 978-3-330-76841-3.

Publisher:
Sciencia Scripts
is a trademark of
Dodo Books Indian Ocean Ltd. and OmniScriptum S.R.L publishing group

120 High Road, East Finchley, London, N2 9ED, United Kingdom
Str. Armeneasca 28/1, office 1, Chisinau MD-2012, Republic of Moldova, Europe
Managing Directors: Ieva Konstantinova, Victoria Ursu
info@omniscriptum.com

Printed at: see last page
**ISBN: 978-620-8-59701-6**

# SUMMARY

# 1 INTRODUCTION

Water pollution is one of today's most serious ecological problems. The sources of water pollution are mainly human activity in the form of domestic and industrial waste (NOGUEIRA; JARDIM, 1998; TERRA et al, 2008), the latter being the main culprits in the contamination of aquatic environments by metals (HOLT, 2000). Once released into the environment, the canons of some heavy metals can be accumulated by organisms and transferred through the food chain, even if present in low concentrations in effluents, as they are not biodegradable and are easily incorporated, becoming dangerously concentrated in fish and other animals, such as molluscs, which are subsequently consumed by humans (BRUN; DIETZ, 1995; HEATH, 1995).

Heavy metals derive from numerous sources, which can be classified as urban-industrial aerosols, liquid and solid human and animal waste, mining and industries, agricultural pesticides and soil source material (NASCENTES, 2006).

The term trace metals includes elements that are needed by organisms in small quantities, such as chromium, copper, nickel and vanadium, as well as those that are not essential and are toxic, even in relatively low concentrations, such as cadmium, mercury and lead (REZENDE, 2009).

Substances whose concentration increases progressively along the ecological food chain undergo the process of biomagnification. Metallic elements are associated with pollution and toxicity and in high concentrations are extremely toxic to life, posing greater environmental risks due to their intense use and wide distribution (BAIRD, 2002).

If these metals are present in relatively high concentrations, they can compromise the entire trophic chain. Considering the bioaccumulative and chronic effects that a metal can have on the planktonic community, microalgae, being the primary producers in the aquatic food chain, are the first to be affected by heavy metal pollution (TONIETTO, 2006). Thus, microalgae are considered to be the entry point of the contaminant into the food web, when the contamination does not occur in lethal doses.

Fish can accumulate copper in various organs and tissues directly, from the water that is filtered through the gills, or indirectly, through food, and this accumulation can result in alterations in the physiological and biochemical processes of the organism. In order to mitigate the effects caused by the presence of metals (HEATH, 1995), organisms can, for example, hypersecret mucus from the gills and skin (TAO et al, 2002) in an attempt to isolate the contaminant.

The concentrations of most heavy metals found in drinking water are usually small and do not cause direct problems for human health. The amounts of metals ingested through our diet are usually much more important than the amounts attributed to the water we drink. However, commonly consumed fish containing heavy metals usually originate from waters with low concentrations of trace elements (BAIRD, 2002).

All metals can be solubilized by water and can cause damage to health depending on the amount ingested, their toxicity or their carcinogenic, mutagenic or teratogenic potential. An organism can exhibit two basic types of behavior in relation to metals: either it is sensitive to the toxic action of a given metal or it is not sensitive, but bioaccumulates it, potentiating its harmful effect along the food chain, putting organisms at the top of this chain at risk (BRAGA et al, 2005).

The hydrographic basin of the Jacarepaguà lowland has a surface area of around 300 km$^2$, covering 18 neighborhoods within the urban area. The basin is made more complex by the presence of around ten inhabited islands (MAIA et al apud FERNANDEZ, 1994). It receives a huge input of nutrients and pollutants due to anthropogenic impacts on its rivers. Within this basin is the Tijuca-Jacarepaguà-Marapendi lagoon system, which is the target of domestic and industrial pollution, making fishing a marginal activity.

The environmental impact study report for the environmental recovery project for the Jacarepaguà macrobasin (JAC, 1998) describes the values for the parameters of temperature, dissolved oxygen and the concentrations of lead, cadmium and zinc in the sediment of the Jacarepaguà Lagoon (table 1). Thus, the sampling point (JAC18) near the Sernambetiba channel showed lower trace element values than the other collection points.

Table 1 - Annual averages, by collection station, of some parametres analyzed in Jacarepaguâ Lagoon.

| Collection point | Temperature ($^0$C) | Dissolved oxygen (mg/L) | Lead (μg/g) | Cadmium (μg/g) | Zinc (μg/g) |
|---|---|---|---|---|---|
| J AC16 | 26,61 | 6,42 | 31,85 | 1,40 | 167,30 |
| J AC17 | 26,82 | 8,75 | 18,03 | 1,05 | 104,65 |
| J AC18 | 26,70 | 11,58 | 3,50 | 0,35 | 31,15 |
| J AC19 | 26,82 | 14,57 | 9,10 | 0,70 | 58,98 |
| JAC20 | 26,63 | 10,52 | 6,65 | 0,53 | 40,95 |

Source: JAC, 1998

Fernandes (1997) studied the sediment of the Jacarepaguâ Lagoon and observed high values of heavy metals, with a zinc concentration of 408 μg.g$^{-1}$ and a lead concentration of 75 μg.g$^{-1}$. Lemes et al (2003) compared metal values in the sediment of the Mogi-Guaçu and Pardo river basins with those of the Jacarepaguâ Lagoon and concluded that the lagoon had higher metal values in the sediment than those found in the Mogi-Guaçu and Pardo rivers. Thus, there is contamination of the sediment of the Jacarepaguâ Lagoon, but studies on the contamination of water and fish by heavy metals have not been found in the literature.

Sardinella Brasilensis, popularly known as sardines, is an economically important species, since in terms of volume it accounts for approximately 25% of the fish produced in Brazil (ACQUA FORUM, 2007). The sardine is a species of fish that stands out for its appreciable consumption and lower commercial price than other fish and for its shoals with large numbers of individuals.

Thus, the sardine is one of the species that inhabits the Jacarepaguâ Lagoon, being intensely consumed by the local population and having no studies on its contamination by trace elements. Strictly speaking, there is not enough information on the quality of the sardines consumed in the region, and the same is true of fish in general, which may be posing a risk to public health.

In the Tijuca-Jacarepaguâ-Marapendi lagoon system, this species is important, either because it is a species that occupies the base of the trophic chain, thus representing the passage of contaminants to secondary consumers in the trophic chain, or because it is a source of protein in the local population's diet, through commercialization and consumption by the riverside population.

Since no study of heavy metal contamination in fish in this region has been found in the literature, previous analyses were carried out to choose the elements evaluated. Thus, after carrying out the pilot analysis, the trace elements were selected because they presented values within the detection limit of the method used.

In view of the above, it is important to survey the concentrations of different trace elements in Sardinella Brasilensis, either because of the importance of evaluating the fish that is consumed by the local population, or because this study could serve as a reference value for future studies in the region.

The main objective of this study was to evaluate the levels of copper, zinc, lead and aluminum in the species *Sardinella brasiliensis* (sardine), which inhabits the outlet of the Sernambetiba Canal, Barra da Tijuca, Rio de Janeiro (RJ).

The specific objective is to analyze the concentrations of these elements in the muscle and viscera *of Sardinella brasiliensis.*

### 1.1 Trace elements, heavy metals or toxic metals?

In the scientific literature, different terms can be found: trace elements, heavy metals or even toxic metals. These terms are often used to describe elements, usually with a high relative density, which have the capacity to accumulate in the biota and which are toxic to organisms above normal levels.

The definitions of the term heavy metal vary according to the branch of science that deals with it. However, each name presents a problem (NASCENTES, 2006). Trace elements are chemical elements (metals and some semi-metals) that have a density greater than 5 g/cm$^3$ (ESTEVES, 1998). Most of these have proven toxic potential, i.e. they are capable of inducing harmful effects on living organisms exposed to them (DALLINGER; RAIMBOW apud DOMINGOS, 2006).

However, the term heavy metal is often used to describe chemical elements that contaminate the environment, causing different types of damage to biota, and can be metals, semi-metals and even non-metals such as selenium. The main chemical elements included in this concept are: aluminum, antimony, arsenic, cadmium, lead, copper, cobalt, chromium, iron, manganese, mercury, molybdenum, nickel, selenium and zinc (TSUTIYA, 1999; PINO, 2005; NASCENTES, 2006).

However, such metals are toxic above a certain amount, and when within normal standards, some are essential elements. For example, the term includes trace elements that are needed by organisms in small quantities, such as chromium, copper, nickel, vanadium and zinc, as well as those that are not essential and are always toxic, even in relatively low concentrations, such as cadmium, mercury and lead.

Figure 1 shows the correlation between the development of living organisms and the concentration of essential (A) and non-essential (B) metals, where the former has a deficient, optimal, toxic and then lethal concentration and the latter a tolerable, toxic and lethal concentration.

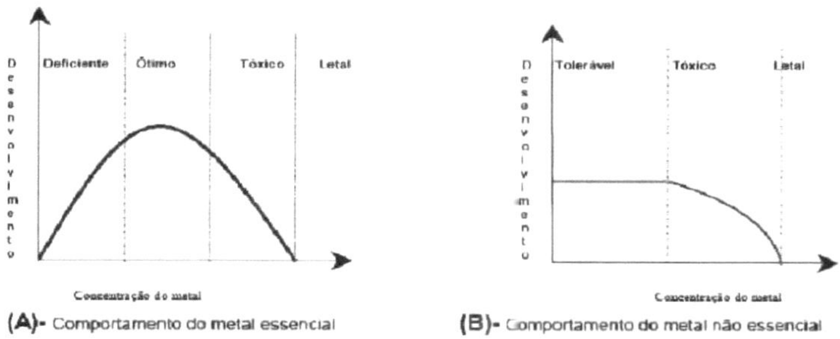


Figure 1 - Correlation between the development of living organisms and metal concentration

Legend: (A) Essential metal behavior (B) Non-essential metal behavior

Source: OLIVEIRA, 2006.

In general, trace elements are metallic elements that are associated with pollution and toxicity and which, in high concentrations, are extremely toxic to life, posing greater environmental risks due to their intense use and wide distribution (BAIRD, 2002).

Thus, they are chemical elements that generally occur in nature in small concentrations, in the order of parts per billion (ppb) to parts per million (ppm). This name is currently the most widely accepted, due to the fact that there is no distinction between metals and non-metals and they occur in low concentrations (ESTEVES, 1998; NASCENTES, 2006). Thus, the term trace elements presents a broader approach; however, the use of other terms should also be considered.

In this sense, the terms heavy metal, trace element and toxic metal are considered synonymous.

### 1.2 Trace elements

Trace elements, from natural or anthropogenic sources, are continuously discharged into rivers, which causes serious problems due to their toxicity, long persistence, bioaccumulation and biomagnification in food (mercury) (ASUQUO et al, 1999; ASUQUO et al, 2004). These react with macromolecules and ligands present in cell membranes, which gives them the property of bioaccumulation (TAVARES, 1992).

Metals are believed to be the most toxic agents known to man. Around 2000 BC, large quantities of lead were obtained from ores as a by-product of the smelting of silver and this was probably the beginning of the use of this metal by man. Industrial activity has significantly reduced the permanence of these metals in ores, as well as altering the distribution of these elements on the planet (LACERDA, 1992).

Trace elements are found in small quantities in aquatic ecosystems, but their levels have been increased by industry, agriculture and mining activities. Their accumulation in fish can be considered an important warning sign for both fish and human health (TURAN et al, 2009). Among foodstuffs, fish and its derivatives stand out as one of the main sources of exposure and contamination by trace elements for humans (WHO, 2000; WIENER; SPRY, 1996).

Trace element concentrations can be considered risk factors for some diseases (KENNISH apud SANTOS, 2009; HEATH, 1995). Some trace elements such as cadmium and lead, in small concentrations, can impair kidney function and cause chronic symptoms of toxicity, including kidney paralysis, reduced reproductive capacity, hypertension, tumors and liver dysfunction (TARLEY, 2001).

The degree of contamination by trace elements depends on the type of pollutant, the species of fish, the collection site, the trophic level and the feeding method (ASUQUO et al, 2004; TERRA et al, 2008). The increase in trophic level influences the concentration of the contaminant, with a few exceptions. Each heavy metal occurs in different concentration ranges, due to the different toxicity of the elements (TERRA et al, 2008; TURAN et al, 2009; TÜRKMEN et al, 2009; TRIEBSKORN et al, 2008; CANLI, ATLI; 2003). The mathematical balance of the concentration of trace elements, resulting from the difference between the amount that enters the body and the amount that is excreted, helps to determine the toxicity of these elements. These elements can be accumulated by organisms through different routes, including respiration, adsorption and ingestion (ZHOU et al, 2001; BORDAJANDI et al, 2003; PTASHYNSKI et al, 2002).

The collection site is directly related to the amount of metal accumulated in the environment, which may or may not become available to the biota. This relationship has been observed in several studies (FERNANDES, 1997; TRIEBSKORN et al, 2008; BAHNASAWY, 2009; PORTO; ETHUR, 2009).

According to their incorporation into the aquatic system list, metals can be classified according to their availability. They can be available, potentially available or unavailable. They are available when dissolved in surface and interstitial waters or adsorbed to the mineral or organic solid phase. Potentially available when they are complexed with organic matter in the form of sulphides, co-precipitated with iron and manganese oxides in the form of carbonates or hydroxides and unavailable when they are in the form of silicates or form the crystalline structure of primary and secondary minerals (COSTA, 2007).

Toxic substances enter an organism's cells through water and ingested food. Inside the cell, these substances can: associate with proteins, such as metalloproteins, with a detoxification function; be transported to the mitochondria; accumulated by the isosomes; transferred to the endoplasmic reticulum or pass through the cell membrane into the blood (hemolymph), thus being transported throughout the animal's body (ZAGATTO, 2006).

According to Chagas (2008), the speed of excretion depends on: the size of the organism, the lipid content (organisms with a high lipid content tend to accumulate more fat-soluble substances), the metabolism (including cellular and molecular) of the organism and the characteristics of the contaminant.

Monitoring using fish helps to determine the quality of the aquatic ecosystem (ADAMS, 2002), so fish samples have been considered among the most indicative factors for estimating trace element pollution in aquatic systems (HAS-SCHON et al, 2006). The concentration of trace elements in fish tissues reflects past and present exposure (CANLI & ATLI, 2003; HENRY et al, 2004), due to the retention capacity and life span of these organisms. Several studies have analyzed the concentration of trace elements in fish. Table 2 shows the results of some studies which evaluated, among other parameters, the trace elements studied in this work in species of the same trophic level as *Sardinella brasilliensis* in areas with different degrees of contamination.

Table 2 - Examples of studies that have analyzed, among other elements, the concentration of copper, aluminum, lead and zinc in fish

| Reference | Year | Species | Tissue analyzed | Cu (mg/L) | Al (mg/L) | Pb (mg/L) | Zn (mg/L) |
|---|---|---|---|---|---|---|---|
| TURAN, C. et al | 2009 | *Engraulis encrasicholus* (anchovy) | Muscle | * | 24, 75 ±4,79 | 0,055 ±0,01 | 14,06 ± ± 1,967 |
| | | | Gonads | * | * | * | * |
| TURKMEN, A. et al | 2009 | *Belone belone* (catfish) | Muscle | 0.8± 0.13 | * | 0.48± 0.13 | 15,0 ± 1,58 |
| | | | Gonads | 1,41 ± 033 | * | 1.82± 0.81 | 23,9 ± 6,58 |
| BAHNASA WY, M. et al | 2009 | *Mugil cephalus* | Muscle | 4.61± 0.62 | * | 9.71± 0.86 | 22.24 0± 6.63 |

| | | | Gonads | * | * | * | * |
|---|---|---|---|---|---|---|---|
| CANLI, M. & ATLI₁ G. | 2003 | *Sardina Pilchardus* (Sardine) | Muscle | 4.17± 0.58 | * | 5.57 ± 1.03 | 34,58 ± 8,64 |
| | | | Gonads | 29.26± 13.1 | * | 39.43± 17.9 | 73.22± 19.84 |

* trace elements not analyzed

The toxicity of chemical substances is affected by factors such as solubility, degree of ionization, route of introduction, storage sites, biotransformation and elimination. In soil and in the bottom sediment of water bodies, the transport, transformation and biological effects of organic molecules depend greatly on their retention in the organic-mineral phase (ARAUCO et al, 2005).

In an aquatic environment, there are four abiotic compartments: suspended particulate matter, sediment, surface water and interstitial water. These compartments have strong interactions with each other and adsorption/desorption and co-precipitation processes can occur between suspended material and contaminants in solution (OLIVEIRA, 2006).

Aquatic complexing agents can be classified as simple ligands (ions) and amino acids, compounds with homologous groups (e.g. humic substances, proteins, metal oxides). Due to the high oxygen content found in the structure of aquatic humic substances, they have an exceptional capacity for complexing metals. This is due to the properties of water macromolecules, such as: the chemical nature of the coordination sites in contrast to simple ligands; the electronic environment of the coordination sites: within a given macromolecule the groups can originate from different fragmentations (e.g. aliphatic chain, aromatic rings), so they can exert different electronic effects and steric hindrance of the sulphur sites (ROCHA, 2003).

The oxygen, nitrogen and sulphur groups are the main ones responsible for the metal complexing characteristics associated with aquatic water substances. Oxygen is found in the form of carboxylic and phenolic groups, while the relatively low concentrations of nitrogen and sulphur are distributed in peptide, sulphonic and thiol groups. These groups have greater binding strength for some metals than oxygenated groups (ROCHA, 2003).

Sediments can act as possible pollution sources for heavy metals, which are not permanently fixed by them, but can be released back into the water column through variations in the physical and chemical conditions of the water, such as changes in pH, redox potential or the presence of organic chelating agents. Factors such as particle size and the degree of crystallinity of clay minerals influence the sorption and retention of contaminants on the surface of the particles. The observed tendency is that when the particle size decreases, the concentrations of nutrients and contaminants increase in the bottom sediment (LEMES et al, 2003). Thus, smaller particles have higher concentrations of contaminants, due to their increased surface area, and because of their size they are easier for biota to ingest.

In continental aquatic ecosystems, trace elements can be in ionic, complexed (mainly with organic molecules) and particulate form (mainly as a component of detritus and biomass). These are also the ways in which these elements reach water bodies. Trace elements can reach the aquatic environment by transport downstream of the site, by transport upstream of the site or by air flows on the free surface. After entering the compartment, the metal can be adsorbed by organic matter, which after increasing in size flocculates and sediments. It is then deposited at the bottom, where the material consolidates. If consolidation doesn't occur, the metal can be resuspended in the water column and exported to the ocean or become available to the biota. However, even after being consolidated, the material can be bioturbated by benthic organisms and undergo the adsorption or desorption process again. These processes are described in figure 2 (COSTA, 2008).

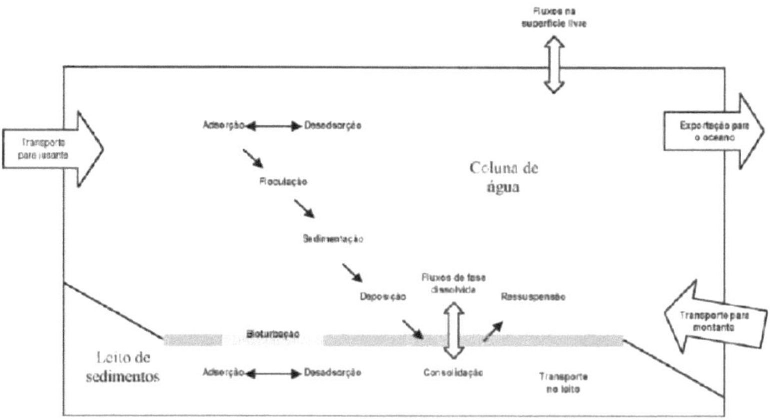


Figure 2 - Forms of adsorption and desorption of metals in estuaries Source: COSTA, 2008.

The mobilization of metals is mainly caused by some changes in the water, according to LEMES ET al (2003):

1 - High salt concentration, whereby alkaline and alkaline earth metal cations can compete with the metal ions sorbed on the solid particles.

2 - Changes in the redox condition, usually associated with a decrease in oxygen potential due to eutrophication, cause iron and manganese hydroxides to be partially or completely dissolved, as well as releasing part of the heavy metals incorporated or sorbed.

3 - Drop in pH, dissolving carbonates and hydroxides, as well as modifying the desorption of metallic cations due to competition with H ions .+

4 - Increased use of natural and synthetic agents, which can form highly stable soluble complexes with metals that are sorbed to solid particles.

In aquatic systems, metals are present in various compartments or reservoirs. A simplified description of the position of metals in aquatic ecosystems is based on four abiotic and three biotic reservoirs. Deposited sediments, interstitial waters, suspended particulate matter and surface water can be distinguished as abiotic reservoirs. Under reducing conditions, deposited sediments can provide interstitial water with a high concentration of nutrients and heavy metals, which can influence the composition of surface water through the process of diffusion, consolidation and erosion, as shown in figure 3. The contribution of the four abiotic reservoirs to the load of heavy metals for aquatic organisms depends on their bioavailability in their habitat (OLIVEIRA, 2006).

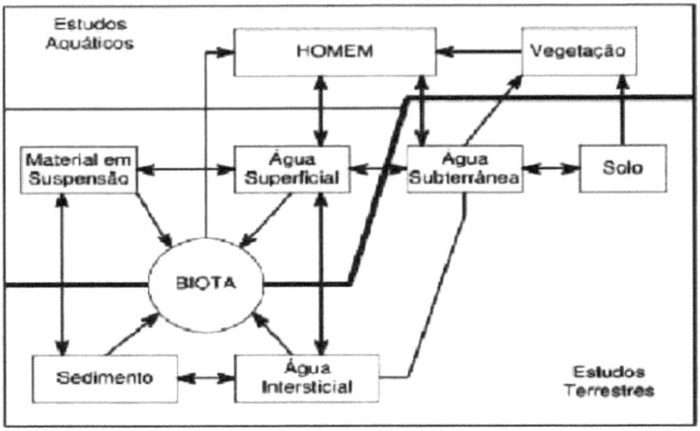


Figure 3 - Simplified description of the metal cycle in the environment Source: OLIVEIRA, 2006.

Sediment, natural particulate material, can normally be transported and deposited at the bottom of aquatic ecosystems, including particles of organic material and water in the interstitial particle phase (ARAUCO, 2005). Some physico-chemical properties of water can affect the rate of adsorption of substances onto sediments, such as temperature, pH, concentration of suspended material, hardness, alkalinity, flow, depth and size of the particles present.

One characteristic of metals that makes them so important as (functional and structural) components of living beings is their propensity to lose electrons easily, forming positively charged ions, which tend to be soluble in biological fluids. It is in their cationic form that metals perform their main biological functions. While metal ions are deficient in electrons, biomolecules such as proteins and DNA are rich in electrons. The attraction between these charge oppositions leads to a general tendency for "metal ions to interact with biological molecules" (BENITE et al, 2007).

### 1.3 Biota

Biota is the set of living beings in an ecosystem, which includes flora and fauna. In aquatic ecosystems, fish reflect the effects of numerous interactions between biotic and abiotic factors. Thus, alterations occurring in fish are considered important indicators of pollution by toxic agents

(RASHED, 2001). The bioaccumulation of heavy metals in fish is evident, even when these contaminants are found in the water in barely detectable concentrations (GOBBI, 2007).

All trace elements are potentially harmful to aquatic organisms at some level of exposure and absorption (TURAN et al, 2009), which depends on the dose and chemical form.

Because metal pollution in aquatic environments can be hazardous to human health it is necessary to understand and control the levels of contamination in fish (LACERDA; GONÇALVES, 2001). Since fish are the last stage in the food chain of aquatic systems, the concentration of trace elements in many fish species has been determined and correlated with the metals present in the aquatic system (TARLEY, 2001).

The study of fish species is important to determine whether they can be used as quality indicators in aquatic ecosystems (TERRA et al, 2008). Biomonitors are important for assessing the bioavailability of trace elements in aquatic systems. Fish are good indicators for monitoring long-term metal accumulation (TÜRKMEN; CIMINLI, 2007; TÜRKMEN et al, 2009).

According to Asuquo et al (2004), fish that ingest pollutants, directly or indirectly, whether from water, debris (food) or sediment, tend to develop chronic diseases. Fish normally resist high exposure to pollutants and develop a detoxification system by sequestering the pollutant, which transforms the metal into a relatively inert form that is less toxic to the organism. Stabilization does not reduce the quantity of the pollutant, but alters its chemical properties, facilitating the sequestration or adsorption of the contaminant, thus reducing the risks (CORRER et al, 2007).

In the normal metabolism of fish, essential metals are taken up from water and food. Different concentrations of trace elements in different fish species may be the result of different ecological needs, metabolism and feeding patterns. It is therefore important to understand the relationship between the size of the animal and the concentration of essential and non-essential metals (CANLI; ATLI, 2003).

Environmental pollution is one of the most important factors in ecological limitation, which corresponds to any factor that tends to reduce the growth potential of an ecosystem, with toxic metals being the main group of pollutants with harmful effects on organisms. Due to anthropogenic interference in nature, pollution and contamination of the environment are a reflection of the load of effluents released into the air, water and soil (ZAGATTO, 2006).

Fish, for the most part, do not live in close contact with the sediment and therefore do not, in principle, have the same significance as benthic invertebrates in assessing the toxicity of sediments. However, many contaminants present in sediments can be diffused into the water column by the constant activity of benthic organisms and also after events such as heavy rains and winds, dredging and turbulence caused by boats.

Fish can also ingest sediment particles whose contaminants may be made available by the acidic environment of the digestive tract, and fish that are not necessarily bottom-dwelling can prey on contaminated organisms that pass between the benthos and the necton.

Although fish are less sensitive than invertebrates to most toxic agents, the standards for toxicity tests recommend the use of organisms from all trophic levels. In addition, fish represent some factors that facilitate dissemination:

1. It is an important source of food for all peoples;
2. They are almost always visible, so the death toll causes astonishment and concern among people, even those who are not sensitive to environmental causes;
3. Among essentially aquatic organisms, they occupy high trophic levels in the food chain and are therefore likely to accumulate high levels of substances through biomagnification;
4. Their organ systems are much closer histologically and physiologically to those of humans (and other vertebrates) than to those of invertebrates, allowing for more reliable extrapolations.

Fish mortality due to metal poisoning is rarely observed in adult organisms, except in highly contaminated samples. In these, however, genetic, histological, biochemical, physiological and behavioral changes can be observed (MELETTI apud MELLETI, 2003).

Xenobiotics (from the Greek, xenos= foreign) are chemical compounds that are foreign to an organism or biological system. It can be found in an organism, but is not normally produced or expected to exist in that organism. The term is also applied to substances present in much higher concentrations than normal. Xenobiotics act on living organisms in different ways and at different levels.

According to Bressola (2007), often a given xenobiotic on its own does not cause adverse effects on the biota, but when it interacts with other substances, it can produce more toxic derivatives, causing serious damage to the entire ecosystem. Although they are not exactly endocrine disruptors, metals such as aluminum interfere negatively with the integrity of sex steroid hormones in fish (CORREIA, 2008).

Therefore, studies of genotoxic, biochemical, histological, physiological and behavioral effects are important, as are studies of bioaccumulation, biotransformation and biodegradation, as well as those that consider the dynamics of populations and communities. A wide variety of damage caused by toxic agents can be observed in organisms, such as histological and morphological changes in bone structure, liver, kidneys and other organs. Fish gills, for example, have sensitive tissues that indicate the presence of irritating materials (MELETTI et al, 2003).

According to Wester and Roghair apud Melleti et al (2003), histopathological studies can reveal specific structural toxic effects in various organs of small fish used experimentally in the laboratory. This type of study is a useful tool in assessing the risk of toxic agents in aquatic environments.

### 1.4 Bioaccumulation

Trace elements have aroused great interest in environmental studies, mainly because they form non-biodegradable compounds, which means that they remain in global biogeochemical cycles. The water cycle is considered to be their main conduit, and they can accumulate in aquatic biota at significantly high levels (CHAGAS, 2008).

The term food chain refers to the sequence in which the beings in a community feed (ODUM, 1983). The process of bioaccumulation of trace elements is one of the ways in which these elements reach high concentrations in the biota, since the levels of some toxic elements increase progressively along the food chain, reaching high levels in predatory organisms that occupy the top of the trophic chain (ZAGATTO, 2008).

A food chain can be defined as the path followed by energy in the ecosystem, from photosynthesizing plants to various organisms that feed on them and serve as food for others, as shown in figure 4.

Figure 4 - Schematic representation of an aquatic food chain Source: LINHARES; GEWANDSZNAJDER, 2003.

Although non-essential elements often pass between organisms and the environment in the same general way as essential elements. Many non-nutrient elements become concentrated in certain tissues, sometimes because of their chemical similarity to specific vital elements, although this concentration can be toxic.

Contamination occurs when a substance foreign to the environment is present. Pollution is the alteration of some environmental quality for which the exposed community is unable to neutralize the negative effects, with some type of risk being identified. According to ROCHA et al (2004), pollution sources can be classified as:

1. point sources: domestic and industrial effluent networks, accidental spills, mining activities, floods, among others;

2. non-point sources: agricultural practices, scattered residences, atmospheric deposition, construction work, soil flooding, among others, and

3. linear sources: highway flooding.

The fraction of the substance that is available for accumulation in living organisms is defined as the bioavailable fraction. Accumulation and, consequently, concentration in the biomass depend on the bioavailability of the substance. Thus, the bioavailability of metals depends on the various chemical forms in which they can occur. Water parameters such as hardness, salinity and pH have a great influence on the speciation of metals and, in general, free ions are more bioavailable (ROCHA et al, 2004).

The toxic effects caused by metals are generally the result of their binding to bionucleophiles. These include amino acids, polypeptides, proteins and enzymes. For example, many heavy metals bind to thiol groups, which often constitute the active sites of many enzymes, crucial for energy supply or oxygen transportation in cells (BENITE et al, 2007). Different ligands such as molecules containing amino, phosphoryl, carbonyl and hydroxyl groups react with heavy metals. The binding of the metal to any of the system's components can affect the entire function (TAVARES, 1992).

Oliveira (2006) uses a simplified classification of the most studied metals in natural waters. This classification is very helpful in understanding the conditions of bioavailability and toxicity of metals, as shown below:

- Elements present mainly in dissolved form: zinc, cadmium, nickel and arsenic;
- Elements present mainly adsorbed to the particles: aluminium, iron, lead and chromium;
- Elements present mainly in the form of colloidal or dissolved organic complexes: copper, lead, mercury, aluminium and iron;
- Elements that are very susceptible to a decrease in pH, in terms of mobility, sedimentation, concentration and speciation: aluminium, manganese, zinc, cadmium, nickel and beryllium;
- Elements present mainly in the form of organometallics formed by biomethylation: mercury, arsenic, selenium, lead and cadmium.

In figure 5, the process of trophic magnification is represented by the increase of the contaminant, represented by the dots, along the trophic chain, which starts with the phytoplankton and continues to the zooplankton, fish and waterfowl. Thus, an increase in the number of dots in the different trophic levels corresponds to an increase in the concentration of the contaminant in the biota.

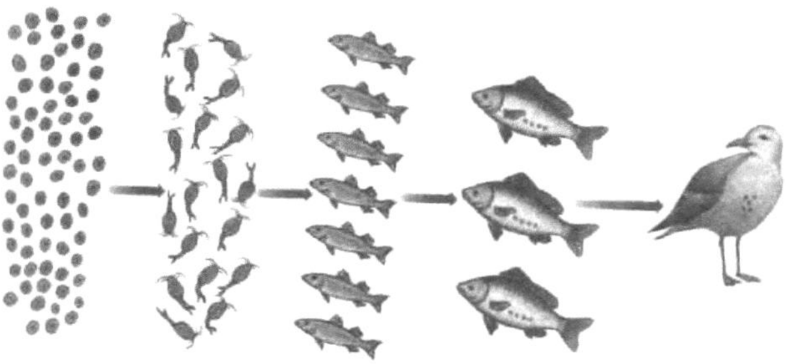

Figure 5 - Schematic representation of a trophic magnification Source: LINHARES; GEWANDSZNAJDER, 2003.

This increased concentration is due to the body's assimilation of these compounds when synthesizing tissues or fats. This increase in the concentration of pollutants along the food chain is called biological amplification, trophic magnification or biomagnification.

**1.5 Specification of metals**

The term speciation analysis is defined as the evaluation of the distribution of metal among various forms or species, i.e. the form in which a metal is present in a given matrix. Speciation analysis can be classified in two ways: physical specification analysis and chemical specification analysis (TONIETTO, 2006).

The ecotoxicity and mobility of heavy metals in the environment strongly depend on their specific chemical form and the nature of the ligands. The speciation and toxicity of metals have been investigated through the use of synthetic and naturally occurring chelating agents, natural waters of different origins and with a variety of metals and aquatic organisms (VAN HULLEBUSCH et al, 2003).

Physical speciation analysis distinguishes the metal between the dissolved, colloidal and particulate fractions, while chemical speciation evaluates the distribution of the metal in various chemical species in solution, considering complexed or non-complexed metals and the distinction between different oxidation states. The study of speciation analysis provides important information from an ecotoxicological point of view, as it allows us to predict a real assessment of the presence of toxic metals in aquatic environments, i.e. those considered bioavailable (TWISS et al, 2001).

There are four important processes for the removal of trace elements: adsorption to iron, manganese and aluminum oxides; adsorption and complexation with particulate organic material; adsorption or utilization by phytoplankton; and export through the main outlets of the system in particulate and dissolved form (OLIVEIRA, 2006).

Figure 6 shows the physical and chemical speciation of metals in the aquatic environment. The metal ion can be associated with organic matter in the particulate fraction or associated with organic compounds and inorganic compounds in the dissolved fraction. Particulate matter does not present toxic forms while inorganic components and hydrated ions are considered bioavailable, i.e. toxic to aquatic biota (TONIETTO, 2006).

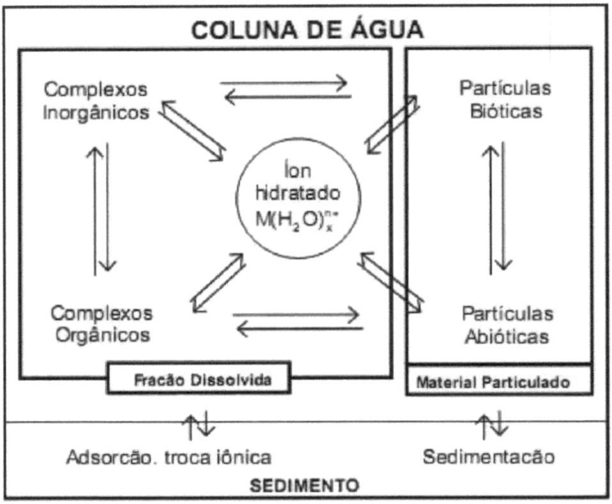


Figure 6 - Schematic representation of the distribution of metals in natural waters

Source: TWISS et al, 2001.

The constituent species of the dissolved fraction is that separated from the natural water sample by filtering it through a 0.45 µm membrane (TEMPLETON et al apud TONIETTO, 2006). In the particulate fraction, metals can be found together with organic or inorganic particles.

According to Sigg apud Tonietto (2006) organic particles larger than 0.45 µm can arise from the decomposition of aquatic organisms or through the degradation products of these organisms. In addition, phytoplankton can interact with metals in this fraction by absorbing the metal or by forming complexes in functional groups present on the surface of the cell membrane (SIGG, 1994). Carboxylic and phosphorylated groups give the cell wall a network of anionic charges in most natural aquatic environments and contribute to the interaction with metal cations.

The adsorption of metals in the particulate fraction can also occur through the association of the metal with inorganic material. In this case, the metal can be adsorbed onto mineral particles such as iron, manganese and aluminum hydroxides as well as clay minerals with silica (SIGG, 1994; OLIVEIRA, 2006). In general, iron hydroxides are the most abundant minerals in natural waters. For this reason, they play a substantial role in the behavior of metals in the water column.

The distribution of a metal species between the suspended particulate material and the dissolved fraction, i.e. its partitioning, plays an important role in the speciation of metals in natural waters.

According to Lu and Allen (2001), knowledge of the conditions that control partitioning makes it possible to assess the extent to which a metal species can be transferred into a potentially bioavailable form. Therefore, understanding this phenomenon is also important for establishing water quality standards for aquatic bodies.

According to Sigg (1994), the main processes that control the partitioning of metals between the solid and aqueous phases are: precipitation, ion exchange and adsorption. The influence of each process on partitioning depends on the type of metal, the characteristics of the suspended material and the composition of the water column.

Gundersen and Steinnes (2003) evaluated the distribution of copper, lead, cadmium and aluminium between the particulate, colloidal and dissolved fractions and observed that, in the case of aluminium, partitioning was governed by precipitation processes, while for the other metals, factors such as pH and organic matter controlled their distribution between the particulate and dissolved fractions.

Lu and Allen (2001) observed that the content of organic matter, both in the dissolved fraction and in the particulate material, governed the partitioning of copper in surface waters. The authors also reported that pH and the concentration of suspended solids are also important aspects in controlling partitioning. The absorption of metals such as zinc and copper can occur directly through water or indirectly through the food chain.

Just as in suspended particulate matter, in the dissolved fraction metal species can interact with a variety of natural and anthropogenic ligands and form stable complexes that are not available to biota. In this reaction, the main constituents responsible for metal complexation are formed by colloidal structures that are counted as dissolved. A good ion exchanger must have mechanical, chemical and thermal stability and extreme insolubility. The affinity of an ion in solution for the exchanger depends basically on the ion's electrical charge, atomic radius and degree of hydration (ROCHA, 2003).

In aquatic environments, copper occurs in both soluble and particulate and colloidal forms, the latter two being the most common. The soluble fraction can contain both the free ion and copper complexed with organic and inorganic ligands. Its speciation in natural waters is determined by the physical, chemical and biological characteristics of the environment. Its transport is mainly in adsorbed form since it is rapidly adsorbed in sediments, resulting in very high residual levels. Copper is bioavailable and can migrate through the environment when it is not bound to organic matter or as an insoluble precipitate (OLIVEIRA, 2006).

Zinc belongs to the group of alkaline earth metals and has the ability to form complexes with ammonia, amines, cyanides, strong and weak bases. When in the sediment, it is mainly associated with Fe and Mn oxides. At low pH (~6.5), this metal complexes with organic matter and suspended minerals (RUSSEL, 1994).

Thus, metals complexed by organic ligands are considered to be non-bioavailable. Despite the low levels of concentration in some natural waters, organic ligands have a high complexing power.

The mobilization of heavy metals present in suspended material and sediment is potentially dangerous for the ecosystem. This mobilization can be caused mainly by four types of chemical changes in the water:

1. High salt concentration at which alkaline and alkaline earth metals can compete with the metal ions adsorbed on solid particles;

2. Change in the redox condition, generally concomitant with the decrease in oxygen potential due to advanced eutrophication.

3. Drop in pH, which can dissolve carbonates and hydroxides, as well as increase the desorption of metal cations due to competition with H ions ;+

4. The increased use of natural and synthetic agents can lead to the formation of highly stable soluble complexes with heavy metals, which are then absorbed into solid particles.

The distribution of trace metals between solution and particulate considered in the study of metal partitioning and speciation:

1. The chemical form of dissolved metal originates from both natural and anthropogenic sources (FORSTNER; SALOMONS apud OLIVEIRA, 2006);

2. The type of interactive process, such as sorption/desorption or controlled precipitation mechanism (SALOMONS apud OLIVEIRA, 2007);

3. Concentration and composition of particulate matter, especially with regard to the active phase on the surface.

In natural aquatic environments, a large part of the components of the colloidal fraction is made up of aquatic humic substances. These substances are the main constituents of dissolved organic matter and are also responsible for the complexation of metals in the soluble phase (ZSOLNAY, 2003). An important characteristic of dissolved organic matter is its ability to form stable complexes with metal ions. It is believed that only a small portion of the total concentration is in the free or bioavailable form (TONIETTO, 2006).

Complexation capacity is defined as the concentration at which a metallic ion can be added to an aquatic medium without its ionic form being the predominant one (FLORENCE et al apud TERRA, 2008). Some authors consider inorganic complexes to be bioavailable or toxic to aquatic organisms (TONIETTO, 2006).

The determination of free metal is important because it is a key parameter in the study of metal reactivity, bioavailability and effects. This form of metal is regulated by the complexation interaction between trace metal in ionic form, ligands and especially ions and particles (TWISS et al, 2001).

During the complexation of a given metal ($M^{m+}$ ) in an aquatic system it can complex with a ligand ($L^{(n-)}$) forming a $ML^{((m-n))}$ complex which can be represented by (1):

$$M^{m+} + L^{n-} \rightarrow ML^{(m-n)} \quad (1)$$

And its formation constant is given by (2):

$$K = [ML^{(m-n)}] / [M^{+}] \times [L^{n-}] \quad (2)$$

where:

$[M^{m+}]$ = concentration of free metal in mol/L

$[L^{n-}]$ = concentration of the free ligand in mol/L

$[ML^{(m-n)}]$ = concentration of the metal and ligand complex in mol/L

In general, it can be said that pH has a strong control on metal speciation, as it determines metallic hydrolysis, polymerization, aggregation, precipitation and proton competition for available binding sites (SMITH, 2002).

It is now known that the free fraction of the metal controls its bioavailability. Thus, for example, the toxicity of copper to fish is not related to the total concentration of the metal, but is strongly associated with its free concentration (OLIVEIRA, 2006).

When two immiscible phases, adsorbate and adsorbent, are brought into contact, the concentration of the adsorbate in one phase tends to be higher at the interface of the adsorbent than inside it. When this occurs, with the consequent accumulation of a substance on the surface of the adsorbent, the phenomenon of adsorption is characterized. The adsorbate, which is the species adsorbed on the surface, is a gas or a solute dissolved in a medium (SHAW; LIMA apud OLIVEIRA, 2006). The forms of adsorption commonly found in environmental studies are physical adsorption, chemical adsorption and ion exchange (O'NEILL apud OLIVEIRA, 2006).

Physical adsorption occurs when an aqueous species is adsorbed onto the external surface of a solid particle by means of weak dipole-dipole interactions, also known as Van der Waals interactions. In this type of interaction there are no three-dimensional molecular arrangements, and it occurs mainly in the adsorption of metals by silicates and organic matter (NAKSHINA apud OLIVEIRA, 2006).

Chemical adsorption is characterized by a direct chemical association between the ions or molecules in solution and the solid phase, for example between metallic oxides and sulphides (GARGIONO apud OLIVEIRA, 2006).

In ion exchange, metals in solution move into the solid phase, replacing ions that were compensating for positive or negative charges in the mineral structure. This mechanism is characteristic of clay minerals (OLIVEIRA, 2006).

Of the hypotheses proposed to describe and interpret the adsorption of metal ions on solid surfaces, the one that contributes most to current knowledge is the theory of adsorption on the surface of hydrated metal oxides. The model assumes that the surface of the oxide functions as a Lewis acid that can receive a pair of electrons from water (STUMM apud OLIVEIRA, 2006).

The hydroxylated oxide (resulting from the dissociation of water) behaves as a weak acid or base (amphoteric character) when the pH of the system varies. The metal present in the water column can interact with the surface. In the presence of ligands, complexation on the surface can be coordinated by the metal.

The adsorption of the metal ion is predicted by the model through factors specific to the adsorbent, such as the density of the adsorption sites and the acidity; or to the adsorbate, through the dissociation constant between the metal and the surface sites (GSCHWEND; WU apud OLIVEIRA, 2006). Although the surface complexation model has been successful as a tool for understanding basic adsorption/desorption interactions, it has some limitations, such as not taking into account the heterogeneity of sites in the natural system and the non-sumulative effect of sorption on multiple adsorbents (OLIVEIRA, 2006).

Tonietto (2006) studied the interaction between microalgae, dissolved organic matter and metals in natural aquatic environments. This study showed that microalgae are responsible for the fate and effects of trace metals in these environments. The importance of microalgae lies in the fact that they have a charged surface and act in the sequestration of metal ions through this surface, and can also absorb and accumulate metal species, which is the entry of the contaminant into the trophic chain.

The studies carried out by Meylan et al. (2003) evaluated the accumulation of copper by algae in response to changes in its species in surface waters and the authors concluded that the intracellular concentration of copper varied as a function of the exchangeable form of the metal, i.e. the sum of the concentrations of the free ion and weak inorganic complexes.

In Brazil, it is only recently that water quality standards for receiving bodies, related to metal species, have been established based on the total or soluble concentration of the metal in the water column (CONAMA, 2005). Before this, water quality standards were based solely on the total concentration of the metal (CONAMA, 1986). The definition of the concentration of the metal in the dissolved fraction as a criterion for assessing water quality is a step forward, since this fraction is more closely linked to the bioavailability of the metal than its total concentration (POTHRO apud ALLEN; HANSEN, 1996).

The metal can be in the water in inorganic or organic form, which in turn can be in soluble form (ion, complexed ion, chelated ion or molecule) or in particulate form (colloidal, precipitated or adsorbed). If the metal is present with other metals or chemicals, there can be a joint action (said to be more than additive), a non-interaction (additive) or an antagonistic action (less than additive).

The physiology of the organism and the forms of metals in the water are altered by factors such as temperature, pH, dissolved oxygen, light and salinity. It also depends on the conditions of the organism in relation to its stage of development, age, size, sex, hunger, activity and adaptation to metals. The sum of these factors leads to an altered behavioral response. In this context, figure 7 summarizes the factors that can influence the toxicity of metals to aquatic organisms.

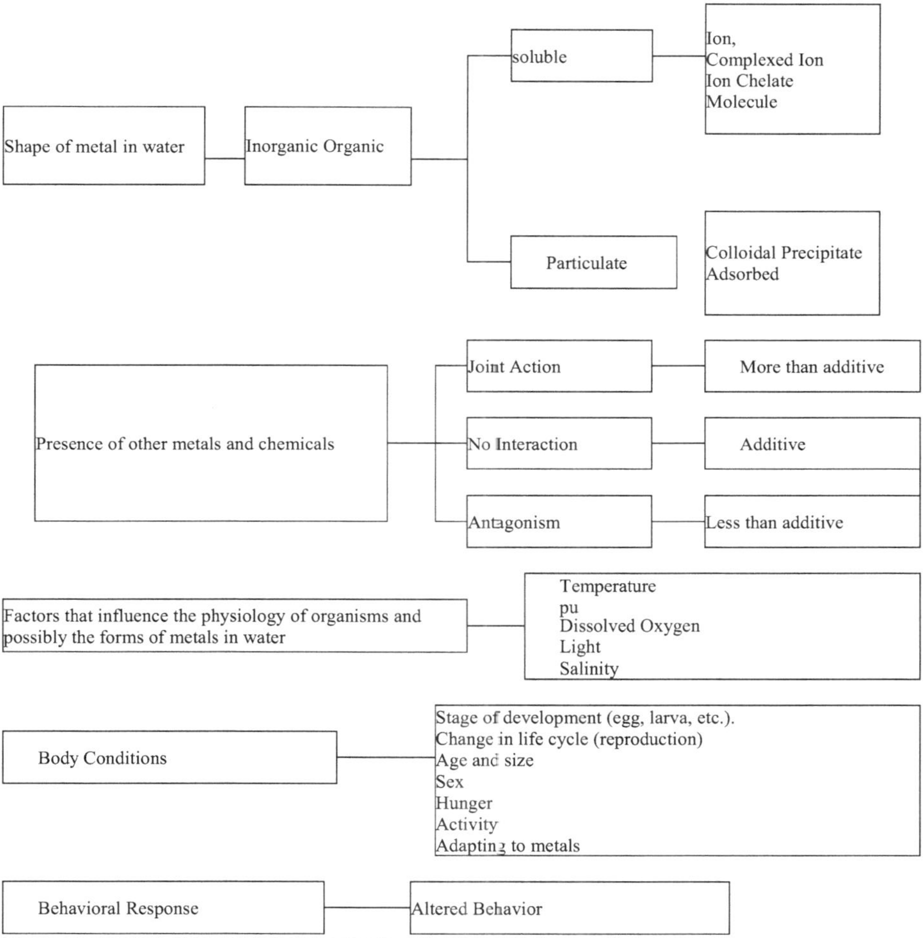

Figure 7 - Factors that can influence the toxicity of metals to aquatic organisms

Source: OLIVEIRA, 2006.

## 1.6 Metal contamination

### 1.6.1 Lead

Lead, even in low concentrations, is foreign to the human body and can interfere with various parts of metabolism and cause poisoning. Chronic lead poisoning is called saturnism or plumbism (MAVROPOULUS, 1999). Lead can be found in the environment in its least toxic form, Pb (0), most toxic, Pb (II), or in an intermediate form of toxicity, Pb (IV). The link between lead and erythrocytes can occur through transporters, such as ion exchange (CALDERON-SALINAS et al, 1996). Lead, in its ionic form, forms precipitates with inorganic phosphate anions.

$Pb_{+}^{2}$ is among the metals that react with the enzymatic thiol group. Ninety-five percent of lead in the human body is transported bound to the hemoglobin of red blood cells and studies of some populations indicate that up to 90% of the total load absorbed in adults is deposited in the bones. Lead crosses the blood brain barrier more in children than in adults, making the nervous system the critical organ in humans, especially children, causing ataxia, convulsions and coma at the level of intoxication, and fatigue, anxiety and memory loss at sub-clinical levels (TAVARES, 1992).

Lead is resistant to atmospheric oxidation and to attack by acids, both hydrochloric and sulphuric, but it is quickly dissolved by nitric acid. Therefore, since the digestive system uses hydrochloric acid during the digestion process, this element becomes difficult to degrade.

There are two classes of lead compounds: the inorganic ones, which are made up of lead salts and oxides, and the organic ones, which are tetraethyl lead and tetramethyl lead. Once absorbed, all inorganic compounds act on the body in the same way. Organic compounds are fat-soluble and can be absorbed through the skin and via the respiratory tract. Because they are fat-soluble, there will be a predominance of nervous disorders.

Once absorbed, lead is not distributed evenly throughout the body. In the blood, circulating lead is almost always associated with erythrocytes and is then distributed to soft tissues (highest concentrations in the liver and kidneys) and minerals (bones and teeth). Aquatic organisms capture and accumulate lead in water and sediment. In fish, lead accumulates mainly in the gills, liver, kidneys and bones.

In recent years, the demand for this contaminant has changed in terms of its type of use. Its use as an anti-knock agent in gasoline and paints has decreased considerably, but its use in industrial processes has increased significantly.

The annual amount of lead dispersed as an atmospheric contaminant is high. In Brazil, there are no known parameters limiting the concentration of lead in the air, nor are there any specific standards for its control in the atmosphere (VANZ et al, 2003).

Soil contamination by lead can occur naturally or geologically - once deposited there, this contaminant can remain indefinitely - as well as through human activities (mining, transportation and industry). Therefore, the lead content in soils varies from region to region: in regions close to high-traffic roads and industries, lead levels are much higher than those found in isolated areas (LARINI, 1993).

Drinking water, with a low pH and low concentrations of dissolved salts, can carry quantities of lead from solder, pipes and fittings, cisterns and reservoirs. When combined with other elements, it creates various compounds. Lead carbonate, cerusite ($PbCO_3$) is commonly found. Lead sulphate ($PbSO_4$) is anglesite; lead chromate ($PbCrO_4$) is crocoisite and lead molybdate ($PbMoO_4$) is wulfenite. Combined with sulphur, lead occurs in the form of sulphide, PbS (galena), which is one of the most abundant lead ores (MAVROPOULOS, 1999).

The World Health Organization (WHO) suggests that the weekly intake tolerance for adults is 3mg of Pb (400 to 450 µg/day). Food (including drinking water and alcoholic beverages) is the major source of lead exposure in the population. Children may have additional exposure from soil and dust (MAVROPOULOS, 1999).

There is also the possibility of lead contaminating food through its migration from containers that contain this food. Oxygen seems to accelerate the release of lead from food containers (CORVALAN et al apud GOBBI, 2007). Lead in ionic form generates precipitates with inorganic phosphate anions.

1.6.2 Zinc

Metallic zinc is used to produce alloys or to galvanize steel structures. This process consists of electrodepositing a thin film of zinc on the parts. Galvanizing is used to protect building structures and the constituent parts of cars and boats from corrosion. This element is also used as an additive in rubbers and paints. One of the most important zinc alloys is bronze, which consists of a mixture of this element and copper. Bronze is more ductile than copper and has a much higher resistance to corrosion. Zinc can also be used as an electrode in dry batteries (UTIMURA; LINGUANOTO apud JESUS, 2004).

Zinc is present in enzymes, participates in the process of hepatic mobilization of vitamin A, acts in growth and sexual maturation, immune functions, among others. It is well known that zinc plays a vital role in animal development. A diet rich in zinc reduces the risk of bleeding and improves wound healing (JESUS, 2004).

This metal, which can complex with HCl to form $ZnCl_2$, is present in metalloproteins in the form $Zn^{2+}$, and is responsible for shielding metals such as cadmium, mercury, copper and others. Zinc is not highly toxic in any chemical form (COTTON et al, 1999).

Zinc is considered to be the twenty-fifth most abundant element in the earth's crust; it occurs in various minerals and in different forms (sulphides or carbonates) and its main use is in galvanizing iron products, providing a corrosion-resistant coating. It is used in batteries and fertilizer, paint, plastics, rubbers, in some cosmetics such as face powders and foundations and pharmaceutical products such as vitamin complexes (GOBBI, 2007).

Zinc is only passively absorbed in the duodenum and jejunum at a rate of 10 to 40% of oral intake. After being released from food, it forms complexes with endogenous and exogenous ligands such as histidine, citric acid and picolinic acid. The survival of the fish is threatened, as it passes into the bloodstream by active transport and combines with albumin and amino acids, inhibiting spawning and the rest of the population, to the detriment of these effects (SORENSEN, 1991).

The low average zinc concentration observed in fish sampled during experiments may be due to this element being poorly concentrated in aquatic biota, even when its concentration in abiotic compartments is high (ADEYEYE et al, 1996).

As reported, excessive zinc interferes with the metabolism of other minerals in the human body, especially iron and copper. After ingesting 2g or more of zinc, symptoms of poisoning appear, which include nausea, vomiting and fever (HENRIQUES et al, 2003).

The main zinc compounds are zinc oxide (ZnO), used in the ceramics and rubber industries and also in the manufacture of paints, and zinc sulphate ($ZnSO_4$), which is used in the textile industry and to enrich soils poor in this element. Zinc chloride is used to preserve wood and as a deodorizer in various fluids (JESUS, 2004).

Zinc does not show variable valence, i.e. it only has oxidation state II (DEMUELENAERE, 2004). It is strongly adsorbed by clay minerals, carbonates and oxides. Due to the relatively high solubility of zinc compounds, precipitation is not the main mechanism for retaining this metal in soils.

Zinc hydrolyzes at a pH greater than 7.7 and the hydrolyzed species are strongly adsorbed by the surface of the soil particles. Zinc forms complexes with inorganic and organic ligands that affect adsorption reactions with the soil surface (COELHO et al apud COELHO et al, 2007).

### 1.6.3 Aluminum

Aluminum is the most abundant metal in the earth's crust (8.13 %) and the third most abundant element after oxygen and silicon. In many countries, aluminum is produced and consumed in large quantities, with Brazil producing around 762,000 tons per year (CASARINI et al, 2001).

Aluminum's toxicity to fish is extremely high, making it the 6th most toxic metal for these animals, after mercury, copper, zinc, cadmium and tin (VUORINEN et al apud PORTO; ETHUR, 2009). In tropical climates, however, little is known about the physiological effects of aluminum on fish, and studies with neotropical species are needed so that the limit concentrations determined in Brazilian legislation for Class 1 waters (0.1 mg/L Al) can be validated (BARCAROLLI; MARTINEZ, 2004).

The pH of the water directly interferes with the toxicity of aluminium on fish, since at a pH of 5.0 the polymerization process of aluminium begins, which is formed by repeated units, called monomers (CORREIA, 2008). The toxicity of aluminum is attributed to the dissolved monomer, the competition between $H^+$ and the cationic forms and the highly toxic polymeric colloids (GENSEMER AND PLAYLE, 1999).

Alonso et al. (2004) evaluated the characteristics of waters from the Guadiamar River basin in Spain and found that changes caused by urbanization, such as variations in pH levels, dissolved organic carbon, dissolved oxygen and suspended solids, affected the speciation of metals in the water column.

The compounds $Al(OH)^{(2+)}$), $Al(OH)_2^+$ , $Al(OH)_3$, and $Al(OH)_4^-$ predominate at pH 5.5, 6.0, 6.2 and around 6.2, respectively (HARRIS, 2007). The forms $Al(OH)_2^+$ and $Al(OH^{)(2+)}$ appear to be the most toxic for most fish species (FAIRMAN; SANZ-MEDEL, 1995).

Aluminum is found in nature mainly as compounds in solid form (salts and hydroxides), but acidification processes (such as acid rain) have increased the mobility of aluminum in aquatic systems (AWWA apud BARBOSA et al, 2000). In addition to aluminum, other contaminants associated with commercial aluminum or raw water can be added to the aquatic system through the discharge of sludge from treatment plants (CORDEIRO apud BARBOSA et al, 2000) and this makes it even more difficult to understand the chemistry and toxicity of aluminum in the aquatic environment.

The chemistry of aluminum in the aqueous environment is extremely complex. Toxicity and bioaccumulation in aquatic organisms depend on the specificity of aluminum in the aquatic system, which is influenced by pH, temperature, concentration with organic ions and duration of exposure. These factors make it extremely difficult to assess the risk of aluminum to ecological integrity because toxicity varies with molecular form and concentration (AWWA, 1991; WREN & STEPHENSON, 1991). For example, dissolved inorganic species, such as aluminium fluorides and hydroxides, are more toxic to plants and fish than natural organic complexes of aluminium, such as when the trace element is complexed with humic substances, which are essentially non-toxic.

1.6.4 Copper

Examples of activities that can contribute to the addition of copper to the environment are: corrosion of copper and brass pipes by acidic water, the use of algaecides, pesticides, fertilizers and fungicides, which are used to preserve wood (CHAGAS, 2008). However, the mining, smelting, electroplating and refining industries are the main sources of contamination. Waste dumps are also responsible for releasing a variety of metals, including copper, into the environment and polluting the soil or water tables. In addition, the manufacture of semi-conductors and other electrical components uses copper, which can cause environmental contamination during the production of these components and devices (CAMPOS et al, 2002).

Copper is a metallic chemical element, which is found in stable compounds with an oxidation state of +1 and +2. The +2 oxidation state is most commonly found and the only one in solution. The monovalent ion $Cu^+$ undergoes disproportionation in water and therefore only exists in insoluble solid compounds or complexes.

Free $Cu^{(2+)\ ions}$ are the most toxic species to unicellular algae, probably because they form complexes with ligands on the membrane surface (TONIETTO, 2006).

In aquatic environments, copper occurs in both soluble and particulate and colloidal forms, the latter two being the most common. The soluble fraction can contain both the free ion and copper complexed with organic and inorganic ligands (OLIVEIRA, 2006).

Due to its high toxicity, copper is one of the metals of greatest interest in the study of the effect of pollution on fish physiology. This metal is often present in freshwater as a result of natural weathering or soil erosion, sewage plants, industrial processes, agricultural use in the form of copper sulphate as a pesticide and also used to control algae in water supply stations and fish farms (MENEZES, 2005).

Copper is considered an essential trace element for the proper functioning of the body (SORENSEN, 1991) and is a co-factor in various enzyme systems. It is required by oxidative enzymes such as superoxide dismutase, catalase, peroxidase and cytochrome oxidase, in the synthesis of hemoglobin, and is transported to the kidneys, heart, central nervous system, bones and stored mainly in the liver (SORENSEN, 1991).

Fish can accumulate copper in various organs and tissues directly, from the water, or indirectly, through food (SEYMORE et al apud MENEZES, 2005), and this accumulation can result in alterations in the physiological and biochemical processes of the organism. Copper causes alterations in the mechanisms of ion regulation, starting with a rapid decrease in plasma electrolytes due to the inhibition of the influx of $Na^+$ and, consequently, the inhibition of the enzyme Na /$K^{++}$ - ATPase in the gills (LI et al apud MENEZES, 2005).

Copper also acts on the permeability of paracellular junctions by causing the displacement of calcium (the main ion responsible for controlling gill permeability), favoring the exit of ions from the cell. The inhibition of the influx of $Na^+$ and $Cl^{(-)\ ions}$, together with the increase in the efflux of $Na^+$ causes a decrease in the concentration of these ions in the plasma. As a consequence, the water present in the plasma flows into the tissues, making the blood concentrated, which is aggravated by the release of erythrocytes due to contraction of the spleen. The high blood viscosity causes the fish's circulatory system to fail and can lead to death (WILSON; TAYLOR apud MENEZES, 2005).

Increasing the hardness of the water reduces the toxicity of this metal, as the concentration of $Ca^{+2}$ affects the toxicity of copper in freshwater fish. This is because $Ca^{+2}$ acts on the permeability of the gills and competes with copper for binding sites in this organ. Thus, calcium is considered an important mediator of copper accumulation and toxicity in fish (HOLLIS et al apud MENEZES, 2005).

Studies on the effect of pH on copper toxicity (TAO et al, 2002) state that a reduction in pH can increase the toxicity of the metal, because in water with an acidic pH, there is an increase in the availability of free copper ions, increasing its lethality. This is because the increase in the concentration of $H^+$ in the medium favors the dissolution of complexes, such as $Cu(OH)_2$, releasing $Cu^{+2}$. Dissolved organic matter is classified as humic or fulvic acids and is considered the main mediator of copper toxicity. Dissolved organic matter binds to the metal and forms complexes, reducing its free form ($Cu^{+2}$), and consequently has a protective effect by altering its speciation in natural waters (PLAYLE et al apud TONIETTO, 2006).

The mechanism of copper toxicity often involves the metal competing for active enzyme sites normally occupied by other metals such as zinc (TONIETTO, 2006). This substitution of one element for another ends up altering vital functions within the organism and can lead to death. Heavy metals in their free form, such as $Cu^{+2}$, are quickly adsorbed on the cell surface of algae or in the gills of fish through interactions between the metal and proteins. The adsorbed metal slowly diffuses through the cell membrane into the interior of the cell, where it can participate in reactions that are harmful to the organism. An excess of this element can lead to disease and toxicity.

The liver is the storage organ and therefore more susceptible to toxicity. Continued uptake of copper, even in low doses, produces considerable accumulation in the liver (WHO apud SANTANA, 2003) and absorbed in toxic doses, it mainly affects the kidneys, liver (causing cirrhosis) and central nervous system (causing depression followed by exc tability).

Copper is mainly transported in adsorbed form, as it is rapidly adsorbed in sediments, resulting in very high residual levels. Adsorption rates vary according to the type of sediment, pH, the presence of competing cations and ligands and Fe and Mn ox des. Desorption also depends on pH, salinity and the presence of natural or artificial chelating compounds. However, copper is bioavailable and can migrate through the environment when it is not bound to organic matter or as an insoluble precipitate (OLIVEIRA, 2006).

In acidic environments, the most bioavailable form of copper is the free cation ($Cu^{+2}$), which is considered the most toxic form for fish; in water with an alkaline pH, the most bioavailable form is the copper hydroxide complex ($Cu(OH_{)(2))}$ (TAO et al, 2002). Thus, the toxicity of free cation is greater than that of copper hydroxide, because free cation can directly replace other cations in the body, whereas ($Cu(OH)_2$) has to dissociate into $Cu^{(+2)}$ and $OH^-$ before it can be assimilated by the body. For this reason, copper toxicity is expected to be higher at acidic pH and lower at alkaline pH.

Tonietto (2006) evaluated the competition between copper and zinc for the available complexation sites in natural water samples. The results showed that zinc does not compete with copper for complexation sites. It is important to consider that a metal is bioavailable to the species in the environment when it is present, or can be converted into its free ionic form, making it capable of being absorbed on a time scale relevant to the deve opment of living organisms and, in this way, being able to affect their life cycles.

### 1.7 Limit values for trace elements in fish for human consumption

Brazil is in twenty-fifth place in world fish production, with an annual production of 820,000 tons of fish (VILA NOVA et al, 2005).

There is no specific legislation that assesses the maximum daily intake limit for all trace elements, specifically in fish. However, the various regulatory bodies set limits for the concentration of these elements in foods, sometimes referred to generally by the term "other foods".

According to Brazilian legislation for "other foods", the maximum permitted limit for zinc in fish is 50 mg/kg (BRASIL, 1965). However, Decree No. 55.871 of March 26, 1965 sets maximum tolerance limits (in parts per million - ppm) for inorganic contaminants that can be found in food, for example in fish: cadmium 1.0 ppm; lead 2.0 ppm; copper 30.0 ppm; chromium 0.1 ppm and; nickel 5.0 ppm. On the other hand, the National Health Surveillance Agency (ANVISA, 2003) only determines the values for cadmium (1 mg/kg) and lead (2 mg/kg).

When it comes to marketing products in MERCOSUR, the trace metal values established are the same as those accepted by ANVISA. On the other hand, the World Health Organization (WHO) only establishes the same parameters as ANVISA and MERCOSUR, only being more restrictive (0.05 mg/kg of Cd and 0.2mg/kg of lead).

Ordinance No. 685 of August 27, 1998 of the National Health Surveillance Secretariat establishes maximum tolerance limits for lead in food under the conditions in which it is consumed. These values correspond to 2 mg/kg of lead in fish and fish products, while the values for zinc, aluminum and copper are not described for fish and fish products. The provisional tolerable daily intake for lead is 3.6 µg^g of body weight, while the provisional tolerable weekly intake is 25 µg^g of body weight, both established by the World Health Organization (SALGADO, 1996). Table 3 compares the different values mentioned.

Table 3 - Limit values established for copper, zinc, lead and aluminum

| Legislation | Parameter (mg/kg) | | | |
|---|---|---|---|---|
| | Cu | Zn | Pb | Al |
| ANVISA | 10 | X | 2 | X |
| MERCOSUR | 10 | X | 2 | X |
| Decree 55.871/1965 | 30 | 50 | 2 | X |
| Ordinance n⁰685/1998 | 10 | X | 2 | X |
| WHO | X | X | 0,2 | X |

X: Values not specified.

### 1.8 Coastal lagoons

The aquatic ecosystem is considered the most susceptible to pollution (BERTOLLETI, 1990), as it is commonly associated with the discharge of domestic, industrial or agricultural effluents and can occur intentionally or accidentally, from natural sources or as a result of human activity (ZAGATTO, 2008).

Coastal lagoons are environments formed by the variation in sea level during successive marine regression events during the Quaternary period (MUEHE & VALENTINE, 1998). In general, they are small and shallow, reflecting an environment of low hydrodynamic energy, with a high residence time of their waters and with their main movements forced by wind waves, tides and storm events

(FERNANDEZ et al, 1994). Because they are regions of intense accumulation of materials, since they are the receiving bodies of the rivers in the drainage basin, they have notable effects from the point of view of geochemical properties, and can be considered favorable sites for studying the evolution of sediment input over several decades, reflecting the history and extent of pollution (RUIZ-FERNANDEZ, 2003).

Due to the process of transport and redistribution of metals in the different phases, apparently low concentrations in the water column and sediments can be potentially available for accumulation in organisms, eventually becoming toxic (SANTOS, 2009).

Lagoons represent 15% of the world's coastal zone and are among the most productive ecosystems in the biosphere. These bodies of water serve various anthropogenic activities related to food, energy, transportation, recreation and urbanism and their natural balance can be easily disturbed, often irreversibly and always accompanied by socio-economic problems (ESTEVES, 1998).

These environments, despite the immense potential they represent for fishing, aquaculture and tourism, have suffered intense and repeated environmental impacts, which have negative effects on their biological productivity (MEDINA et al apud BARROSO et al, 2000).

The growth of urbanization without adequate infrastructure and other anthropogenic activities have completely modified the natural characteristics of various depositional environments, such as coastal lakes and lagoons, affecting their environmental quality due to the discharge of domestic and industrial waste into their water bodies (CYPRIANO et al, 2008).

Fish, the main target of fishing activity, are mobile populations that participate in energy regulation, and their abundance and diversity is a biological indicator of the environmental quality of a lake (YANEZ-ARANCIBIA apud BARROSO et al, 2000; YANEZ-ARANCIBIA & NUGENT apud BARROSO et al, 2000).

Preliminary studies (GONÇALVES, 1999; LACERDA et al, 2001) have shown that the coastal lagoons of the state of Rio de Janeiro are typical examples of the importance of diffuse sources as the main source of contamination of this element in coastal environments, with no point source contributing to the lagoons. However, results obtained for the region have shown high atmospheric deposition rates and relatively high concentrations in the biota of these environments.

The state of Rio de Janeiro has a succession of lagoon systems where fishing is still a very important socio-economic activity, despite the blatant reduction in environmental quality. The Tijuca-Jacarepaguà-Marapendi lagoon system (figure 8) is located in the area of urban expansion and is the target of domestic and industrial pollution, which has made fishing a marginal activity (IBAMA, 2000).

Figure 8 - Schematic map of the Tijuca-Jacarepaguà-Marapendi lagoon system

Source: GOOGLE MAPS, 2010.

Artisanal fishing exists all along the Brazilian coast, and the fishermen who use it operate in areas close to the coast, in coastal bays and lagoons, which are commonly breeding grounds for shrimp and other species of marine fish. These fishermen are generally semi-illiterate and have low purchasing power, but because they are so numerous, they contribute significantly to total fish production (IBAMA, 2000).

Fish production is destined for both subsistence and commerce. Fish with a high added value are preferably destined for sale, while fish with a low sale value or with large shoals are destined for both purposes.

### 1.9 Sardinella brasiliensis

Sardinellas brasiliensis used to swim in large shoals through the seas of Sardinia, an island located in the Mediterranean, from where they got their name. They swam for miles until they spread populations of their species throughout the world's oceans. Over time, they colonized Brazilian waters, forming the species Sardinella brasiliensis, commonly known as the true sardine, which belongs to the class Osteichtyes, the order Clupeiformes and the family Clupeidae. It has been included in the human diet for many centuries.

Sardines are found in greater abundance along the area between the states of Rio de Janeiro (Cabo Sâo Tomé, 22°S) and Santa Catarina a little to the south of Cabo de Santa Marta Grande 28°S (IBAMA, 2009). This species of fish stands out for its appreciable consumption and lower commercial price than the others.

Strictly speaking, there is not enough information on the quality of sardines, and the same is true of fish in general. This situation may be allowing fish to be marketed in an inappropriate way, resulting in an inherent risk to public health (PEREIRA & TENUTA-FILHO, 2005).

The study of feeding habits is one of the main aspects of fish biology. It is an important indicator of ecological relationships and can be used to determine species coexistence strategies, by separating feeding habits by area, time or ontogeny.

S. *brasiliensis is* an omnivorous species, as it preys on zooplankton, but is able to change its feeding strategy to filter-feeding on phytoplankton and microzooplankton (SCHNEIDER; SCHWINGEL, 1999).

# 2 ATOMIC ABSORPTION SPECTROMETRY

This technique aims to study the absorption of radiant energy in the ultraviolet-visible zones by neutral atoms in the gaseous state. In atomic absorption, samples are vaporized at 2,000-8,000 K and decomposed into atoms. The concentrations of atoms present in the vapor are measured by emission or absorption according to the characteristic of the visible light radiation (Harris, 2007). This process is called dissociation, where atoms or molecules are not analyzed, but rather the atomic particles themselves. An ionized gas is called a "plasma" and is considered to be the fourth state of matter. In any state, the change to the next state consumes energy.

Because it is extremely sensitive, this technique allows different elements to be distinguished in a complex sample. Samples are measured in parts per million (µg/g or ppm) to parts per trillion (pg/g). To analyze most constituents, samples must be diluted to reduce concentrations to mg/L. Similarly, trace elements can be pre-concentrated for analysis.

The atomic absorption spectrophotometer makes it possible to determine the presence of trace metals. Used in analytical laboratories, this equipment is used for qualitative and quantitative evaluation. The sample is gasified using a flame. Its atoms absorb part of the radiation energy coming from the method. This reduces the intensity of the light that reaches the monochromator, a device that uses diffraction to separate the different component frequencies of white light. This reduction is specific to each element and can be evaluated on the computer. Figure 9 shows a schematic representation of this process.

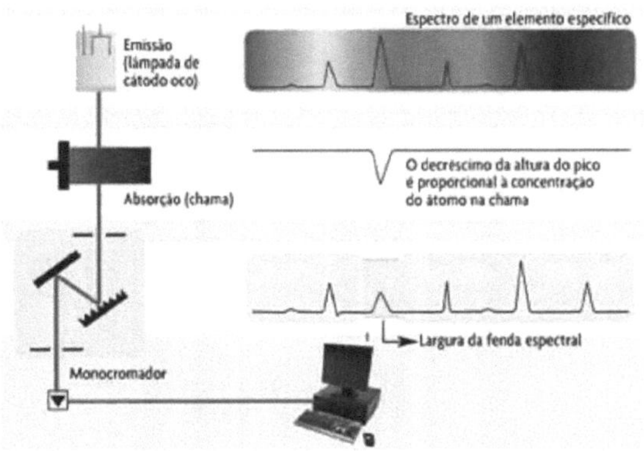


Figure 9: Schematic representation of atomic absorption equipment.

Thus, in an atomic absorption apparatus, the following instrumental systems can be distinguished (Rocha et al, 1954):

1. Emission system - consists of a radiation source that emits the spectrum of the element to be analyzed; the emitted radiation is directed towards the absorbing medium formed by the atoms in the sample;

2. absorption system - the main component of this system is the atomic vapor which will absorb part of the energy emitted by the source. In flame atomic absorption photometry, the absorption system consists of the flame, burner, sprayer, capillaries, tubes or injection system, regulators and pressure gauges to control the pressure of the gases;

3. selection system - refers only to the spectral selection equipment. It includes the optical part for spectral selection (filters, monochromators with prism or mesh) and mechanical accessories (slits, etc.) and,

4. Detection or recording system - Consists of photodetection and measurement systems. This includes all the equipment needed to power the photomultiplier, amplifier, etc.

**2.1** Limitations of atomic absorption spectrometry

As the atomic absorption spectrometry method is a comparative method, the quality of the standards has a considerable influence on accuracy and very pure metals, which by acid attack pass into the form of salts, or high purity pro-analysis products should be used during preparation.

There are two main reasons for chemical interference:

1. the conversion of the sample into atoms may not be quantitative, due to difficulties in melting and vaporizing the salt; and

2. the molecules may not be completely dissociated, or the free atoms may react spontaneously with other atoms or radicals nearby, making them unavailable for absorption, which depends on the stoichiometry of the flame and the affinity of the atoms for others in the flame.

Physical interferences are caused by various physical properties of the solution that affect the measured absorbance value:

1. vapor pressure and surface tension influence the size of the droplet;

2. viscosity influences the speed with which the aerosol reaches the flame by aspiration.

3. Background interference occurs for two fundamental reasons:

4. the scattering of light by solid particles in the flame;

5. the actual absorption by molecules or free radicals coming mainly from the sample matrix.

6. These types of interference cause an increase in the signal and therefore in the absorbance (Gonçalves, 1983). The efficiency and reproducibility of atomization determine the sensitivity, precision and accuracy of the analytical methodology. For this reason, it is recognized as a critical step in this instrumental technique (Luna, 2003).

**2.2 Advantages of atomic absorption spectrometry**

Atomic absorption spectrometry is used to determine approximately seventy chemical elements. The sensitivity of this technique allows chemical elements to be determined in the ppm or ppb range. It also has the following advantages, among others: speed of determination, high selectivity and relatively low cost of the necessary equipment. Other advantages include:

1. use of a simultaneous double beam system, capable of correcting fluctuations in the intensity of the lamp and in the transmission of radiation, without damaging the signal-to-noise ratio;

2. the ability to carry out multi-element determinations using a suitable detector;

3. automated rapid sequential multielemental analysis with flame atomization;

4. absence of "weak lines", since the emission is continuous and the intensity is approximately uniform in the various regions of the UV-visible spectrum;

5. simultaneous correction of the background in the vicinity of the analytical line;

6. the software enables full post-reading signal processing, allowing suitable reference pixels to be selected and integration limits to be set;

7. possibility of storing reference spectra of background absorption caused by molecules with fine structure, for subtraction from the sample spectrum using a least squares algorithm;

8. automatic correction for all continuous events, such as molecular absorption and other broadband phenomena like atomizer emission;

9. calibration range of up to five orders of magnitude;

10. better analytical performance in the determination of trace elements in complex samples (Borges, 2005).

# 3 EXPERIMENTAL PART

## 3.1 Study area

The Jacarepaguâ coastal plain is located in the urban expansion zone of the city of Rio de Janeiro, in the Barra da Tijuca region. The Baixada de Jacarepaguâ basin covers an area of around 300 $km^2$ and includes the neighborhoods of Jacarepaguâ, Anil, Gardènia Azul, Cidade de Deus, Curicica, Freguesia, Pechincha, Taquara, Praça Seca and Tanque, all part of the Jacarepaguâ Administrative Region, and the neighborhoods of Joâ, Barra da Tijuca, Itanhangâ, Camorim, Vargem Grande, Vargem Pequena, Recreio and Grumari, in the Barra da Tijuca Administrative Region.

The basin's watershed is made up of the crests of Pedra da Gàvea, Mesa do Imperador, the Tijuca Massif (Serra dos Pretos Forros, Sâo Francisco, Très Rios, Mattheus, Carioca and the Alto da Boa Vista elevations), Serra do Engenho Velho and Morros do Catonho, Monte Alto, Sâo José and Covanca, continuing through the Pedra Branca Massif (Serras de Santa Bàrbara, Sacarrâo, Nogueira and Quilombo Grumari and Geral de Guaratiba). Thus, its basin is complex, with some rivers flowing into the lagoon system, as well as several indentations and around ten inhabited islands (MAIA et al apud BARROSO et al, 2000).

The basin is formed by the rivers that descend from the slopes of the Tijuca and Pedra Branca massifs and the rocky shield to the north of the lowlands, and by the Tijuca, Camorim, Jacarepaguâ, Marapendi and Lagoinha lagoons. The drainage goes first to the lagoons and then to the sea. Of the total area of the basin, around 180 $km^2$ refer to the surface drained by the rivers.

The Jacarepaguâ Lagoon is a coastal body of water located in the western part of the city of Rio de Janeiro. It is an example of an aquatic ecosystem that suffers environmental impacts due to anthropogenic influences. This lagoon covers an area of 3.7 $km^2$, with an average width of 0.93 km and a length of 4 km, with an average depth of 1.5m. It has a salinity gradient from the Tijuca lagoon to the Atlantic Ocean, with a pH between 6 and 9 and high concentrations of nutrients, mainly ammonia nitrogen and total phosphorus (FERNANDES, 1993; MAGALHAES et al, 2001).

The Jacarepaguà lagoon is home to a large number of rivers and streams that drain the whole of the Jacarepaguà lowlands, running through densely populated areas, as well as areas where many different industries are located (pharmaceutical, chemical, metallurgical, among others) (JAC, 1998). Over the last few years, the system's lagoons have been used to receive domestic and industrial effluents produced in the Baixada de Jacarepaguà, completely changing their main characteristics (CYPRIANO, 2009).

The Tijuca Lagoon, in turn, receives the waters of the Marapendi Lagoon through the canal of the same name, which is about 4 km long. The waters then flow together to its bar via the Joatinga Canal. This hydrographic system has two connections with the sea, one to the east, at the Joatinga Canal, and the other to the west, at the Sernambetiba Canal. Through the former, saltier sea water enters the Tijuca Lagoon and from there into the Marapendi Lagoon, through the Marapendi Canal.

In the case of the Tijuca Lagoon, tidal penetration is attenuated, reaching negligible values at the height of the Camorim Lagoon. The Marapendi Lagoon is located between a narrow strip of beach and the inland lagoons (Tijuca, Camorim and Jacarepaguâ). It is about 10 km long and 350 m wide on average. It is therefore elongated in shape, divided into 7 pocket-like compartments that reduce its capacity for renewal. The lagoon system interacts with the adjacent sea through the aforementioned connecting channels, the Joatinga and Sernambetiba. The exchange of water between the two systems, through the action of the tidal regime, is not enough to renew most of the lagoon waters.

The Sernambetiba Canal, currently closed by rockfill, attenuates the action of the tide, practically preventing its penetration and not renewing its waters. This channel plays a fundamental role in the flow of water from the Pedra Branca Massif and the drainage of the entire western part of the Baixada de Jacarepaguà. The closure of its bar, while on the one hand improving the bathing conditions of the adjacent beaches, contributes to flooding its banks and worsening the quality of its waters.

The Joatinga Canal plays a fundamental role in the exchanges between the Tijuca and Marapendi lagoons and the sea. Due to the large volume of water that passes through the canal each tidal cycle, it is certain that water from the lagoons reaches the adjacent coastal region, altering the natural conditions and damaging their bathing conditions. As a result, Barra da Tijuca beach suffers the most (SEMADS, 2001).

In the surroundings of this lagoon system, the socio-economic contrasts are extreme, as the wealth of condominiums coexists with the misery of masonry shacks, mountains of garbage and colony grass. Domestic and industrial waste containing heavy metals is discharged into the basins that contribute to the Jacarepaguà Lagoon. On the banks of the entire lagoon, there are several sewage dumps coming from the large buildings in Barra da Tijuca and the industrial zone in Jacarepaguà. There have even been reports of illnesses in fishermen living on the islands of the Tijuca lagoon, due to the contamination of fish, as well as anomalies in fish (FERNANDEZ, 1994).

The silting up and landfilling of the banks has led to the reduction of the mangroves, a natural breeding ground for various species of fish and crustaceans. The large amount of manpower needed for the rapid urban growth of Barra da Tijuca has contributed to the slumification of the banks of the entire system. Throughout the lagoon system, wooden caique-type boats are used to catch fish with 30 mm mesh waiting nets and tarrafas. The best quality fish are found near the channel connecting the lagoon to the sea, where they enter with the water at high tide.

Fishing in the lagoon system has been compromised due to reduced productivity as a result of environmental degradation. However, species of economic interest still occur. The Marapendi lagoon, a type of restinga interordoes, is shallow and therefore very fragile and susceptible to disappearance. Dark water with domestic sewage from various canals that pass through residential neighborhoods is constantly discharged, but due to the infiltration of marine waters through the restinga, the water quality in this lagoon is better than in the rest of the system.

Thus, the Jacarepaguà lowland watershed receives a large amount of nutrients and pollutants due to anthropogenic impacts on its rivers. This lagoon complex has two areas of water exchange with the sea. This hydrographic system has two connections with the sea, one through which the saltier sea water enters the Tijuca Lagoon and from there to the Marapendi Lagoon, via the Marapendi Canal, and the other to the west, via the Sernambetiba Canal. However, the first connection with the sea is currently closed by a sand bar, which is mechanically removed from time to time.

Thus, for most of the year, the water is only renewed by the Sernambetiba Canal, where it ends up presenting the best quality conditions in the lagoon complex.

### 3.2 Collection

The samples were collected at the exit of the Canal de Sernambetiba, Barra da Tijuca, Rio de Janeiro (RJ), during the months of May, July and October 2009. The samples were collected with the help of a tarrafa by a professional fisherman. Once collected, the samples were duly identified, packed in plastic bags and frozen for the experimental analyses. Figure 10 shows the sample collection point.

Figure 10 - Satellite photo of the Sernambetiba Canal Source: GOOGLE EARTH, 2009.

### 3.3 Material

The following materials were used in the laboratory: 50 mL glass beakers, heating plate, pipettes, test tubes, precision balance, glass rod, Swinex filter paper holder®, Whatman No. 42 cellulose filter paper, standard solutions of lead, aluminum, copper and zinc, Qhemis brand - High purity (table 4) and Varian atomic absorption spectrophotometer, model AAS-240.

Table 4 - Specification of the standard solutions of lead, aluminum, copper and zinc, brand Qhemis - High purity used for the calibration of the atomic absorption spectrophotometer.

| Métal | Reference | Lot | Manufacturing | Validity |
|---|---|---|---|---|
| Lead | PAPB 1,000 0125 | F08J166L | 12/08 | 24 months |
| Aluminum | PAAL 1,000 0125 | F08J260L | 12/08 | 24 months |
| Copper | PACU 1,000 0125 | F08L205L | 12/08 | 24 months |
| Zinc | PAZN 1.000 - 0125 | F09A098B | 02/09 | 24 months |

### 3.4 Reagents

SIGMA-ALDRICH nitric acid and sulfuric acid, all ultrapure. The reference material Dorm-3® (standard material composed of fish protein impregnated with trace metals, supplied by the National Research Council of Canada) was used to check the precision and accuracy of the method. Table 5 shows the reference values of this material for the different trace elements.

Table 5 - Values of the Dorm-3 reference material® for the elements analyzed

| **Elements - trace** | **Concentration (mg/kg)** |
|---|---|
| Copper | 15.50± 0.63 |
| Lead | 0.395± 0.050 |
| Zinc | 51.30± 3.1 |

### 3.5 Pilot test

To help choose the elements to be analyzed and adjust the analytical methodology, four previous analyses were carried out, known as pilot tests. In these tests, the following chemical elements were analyzed: copper, chromium, cadmium, nickel, zinc, lead and aluminum.

In total, 16 individuals were analyzed during 4 pilot tests, using different fish masses and sample dilutions. Based on the results obtained, the analytical methodology used in this study was established.

### 3.6 Analytical methodology

Each specimen caught was considered an experimental unit, in order to verify the variation in the degree of contamination between individuals.

In the laboratory, each specimen was measured using a caliper and weighed on a precision scale. The total mass considered in this study was characterized by the mass of the fish with the viscera.

Initially, all the glassware used was soaked for 24 hours in a 10% aqueous $HNO_3$ solution and then in plenty of deionized water. After weighing, the specimens were gutted using a plastic knife and each viscera was weighed. For the laboratory analyses, 1,000 g of muscle tissue were used in duplicate and the total mass of the viscera from each specimen.

The samples for analysis of aluminum, lead, copper and zinc were opened using the wet method as cited by Anvisa and Ial (2005).

After about 1,000 g of the wet sample had been weighed, it was transferred to a beaker in which 10.0 mL of aqua regia (3 $HNO_3$ : 1 HCl) was added and then placed on the hotplate at 100°C.

The sample was opened for approximately 2 hours, until the sample was dry without calcining. After solubilization, the samples were cooled to room temperature and 2.0 mL of concentrated HNO3 was added. The samples were placed back on the plate until dry. The samples were then filtered using a Swinex filter paper holder® and resuspended in distilled water until the solution was 25 mL in a beaker. The samples were stored in the refrigerator until they were read. A Varian AAS-240 atomic absorption spectrophotometer (AAS) was used for the readings. The readings were carried out in duplicate. For each reading cycle, 2 samples of distilled water, called blanks, were used to minimize possible analytical errors.

To certify the values found in the analysis, the following material was used

reference Dorm-3, which is a standard material composed of fish protein impregnated with trace metals, supplied by the National Research Council of Canada.

To calculate the concentration of metal in the sample (total mass, muscle or viscera), the following equation was used:

$$Cp = Cs.Vs/m \quad (3)$$

Where:

Cp = metal concentration in the fish sample (mg/kg)

Cs = metal concentration in the solution (mg/L)

Vs = volume of solution ($25 \times 10^{(-3)}$ L) m = mass of sample (kg)

The values found by the above equation refer to each sample, carried out in duplicate. The final concentration for each individual was calculated using the average of the duplicate values, where the difference between the values for each sample was no greater than 10%.

The total concentration was obtained from the weighted average of the concentrations, according to the following formula:

$$Ct=\frac{[(Pm.Cm)+(Pv.Cv)]}{Pt}$$

Where:

Ct = total concentration

Pm = muscle weight

Cm = muscle concentration

Pv = viscera weight

Cv = concentration in the viscera

Pt = total weight

### 3.7 Statistical analysis

Characteristic clustering is a multivariate statistical method that considers samples and variables as a whole, making it possible to extract complementary information that univariate analysis cannot provide. One of the objectives of using multivariate analysis is to reduce the dimensional representation of the data, organizing it in a structure that makes it easier to visualize the whole set (COELHO et al, 2007).

Cluster analysis is used to classify samples by the values of a set of variables within a group. With this method, samples or variables are grouped according to their similarities. The results are presented in the form of a dendrogram, allowing the distances between the samples or variables to be visualized.

This is a classification technique using algorithms, which organizes the observed data into structures that group together similar subsets. It is a multivariate technique that aims to group data according to the similarities between them, forming groups with homogeneity within the cluster and heterogeneity between them.

The method of unweighted means of paired groups has the distance between two groups calculated by the average of all pairs of items belonging to the groups. This method is used when objects form distinct natural clusters (SALDANHA; BENASSI; BRAGGNOLO, 2008; COELHO et al, 2007).

In order to assess the variations in the values of this study, the following groupings were made:

1. Size, total weight and viscera weight;
2. Size, total weight, viscera weight and concentration of each trace element in viscera and

muscle;

3. Trace element concentration in muscle;

4. Trace element concentration in viscera.

# 4 RESULTS AND DISCUSSION

## 4.1 Analysis certification

The values of the reference material Dorm-3® used to measure the accuracy of the methodology are shown in Table 6. The samples were run in 3 cycles (n = 3) and in triplicate (TURKMEN et al, 2009; PRADIT et al, 2010; VALITUTTO et al, 2006) to validate the results.

Table 6 - Values observed in the reference material (Dorm-3® ) (mg/Kg)

| Sample | Concentration (mg/kg) | |
|---|---|---|
| | Cu | Zn |
| 1 | 13,96 | 49,06 |
| 2 | 16,57 | 53,52 |
| 3 | 14,88 | 50,68 |
| Average | 15,14 | 51,09 |
| Desv Pad | 1,320 | 2,260 |

Table 7 compares the observed values with the expected values of the reference material. In this way, the accuracy of the method has been validated.

Table 7 - Comparison between observed and expected values

| Values | Concentration (mg/kg) | | |
|---|---|---|---|
| | Cu | Zn | Pb |
| Expected | 15,50 ±0,63 | 51,30 ±3,1 | 0.395± 0.050 |
| Observed | 15.14± 1.32 | 51.09± 2.26 | < L.D.* |

* Value below the method's reading detection limit.

## 4.2 Pilot test

A total of 16 individuals were used in the pilot test. In the first sample, 1,000 g of sample were analyzed in duplicate (1 and 1'), while in the second, 2,000 g of sample were analyzed without duplicate (2), due to the low weight of the species. Thus, whenever duplicate analysis was tested, the mass weighed was 1,000 g and in the samples where 2,000 g was used, there was no duplicate.

Specimens 3 and 4 were tested in duplicate (3 and 3'; 4 and 4') and specimens 5, 6 and 7 without duplicate. From the following individuals (8 and 8'; 9 and 9'; 10 and 10') the specimens began to have their viscera analyzed (V8, V9 and V10). The chromium, cadmium and nickel values were below the detection limit of the method, while the aluminum values were below the detection limit only in the muscles.

In the following samples, the samples were diluted to 25.0 mL and nickel, cadmium and chromium were not analyzed, as they had values below the detection limit in the previous samples. The muscles of six individuals (11 and 11'; 12 and 12'; 13 and 13'; 14 and 14'; 15 and 15'; 16 and 16') and their viscera (V11, V12, V13, V14, V15 and V16) were also analyzed. Table 8 summarizes the variables used in the pilot test.

Table 8 - Variables used in the pilot tests

| Individual | Duplicate | Sample mass (g) | Analyzed muscle | Viscera analyzed | Total volume of solution (mL) |
|---|---|---|---|---|---|
| 1 | Yes | 1,000 | Yes | No | 50,0 |
| 2 | No | 2,000 | Yes | No | 50,0 |
| 3 | Yes | 1,000 | Yes | No | 50,0 |
| 4 | Yes | 1,000 | Yes | No | 50,0 |
| 5 | No | 2,000 | Yes | No | 50,0 |
| 6 | No | 2,000 | Yes | No | 50,0 |
| 7 | No | 2,000 | Yes | No | 50,0 |
| 8 | Yes | 1,000 | Yes | Yes | 50,0 |
| 9 | Yes | 1,000 | Yes | Yes | 50,0 |
| 10 | Yes | 1,000 | Yes | Yes | 50,0 |
| 11 | Yes | 1,000 | Yes | Yes | 25,0 |
| 12 | Yes | 1,000 | Yes | Yes | 25,0 |
| 13 | Yes | 1,000 | Yes | Yes | 25,0 |
| 14 | Yes | 1,000 | Yes | Yes | 25,0 |
| 15 | Yes | 1,000 | Yes | Yes | 25,0 |
| 16 | Yes | 1,000 | Yes | Yes | 25,0 |

All the samples showed zinc values above the method's detection limit, except for V5. For the copper analysis, samples 11, 11', 12, 12', V11, V12 and V13 had values above the detection limit. The values for lead and aluminum, on average, were below the method's detection limit, except for samples 3, 4 and 4'. However, lead showed values within the method's detection limit for the viscera samples, except for sample V1 (table 9).

Table 9 - Concentrations of Cu, Cr, Cd, Ni, Zn, Pb and Al obtained in the analysis

| Amosfra | Concentration (mg/kg) | | | | | | |
|---|---|---|---|---|---|---|---|
| | Cu | Cr | Cd | Ni | Zn | Pb | Al |
| 1 | 0,800 | 0,425 | <LD | <LD | 12,962 | 3,75 | - |
| 1' | 0,900 | 0,375 | <LD | <LD | 16,257 | 2,75 | - |
| 2 | 1,550 | 2,125 | <LD | <LD | 22,562 | 5,25 | - |
| 3 | 0,200 | <LD | <LD | <LD | 26,205 | <LD | <LD |
| 3' | 6,050 | <LD | <LD | <LD | 35,540 | <LD | <LD |
| 4 | <LD | <LD | <LD | <LD | 50,755 | <LD | <LD |
| 4' | <LD | <LD | <LD | <LD | 15,755 | <LD | <LD |
| 5 | <LD | <LD | <LD | <LD | 23,110 | <LD | <LD |
| 6' | <LD | <LD | <LD | <LD | 70,085 | <LD | <LD |
| 7 | <LD | <LD | <LD | <LD | <LD | <LD | <LD |

| | | | | | | | |
|---|---|---|---|---|---|---|---|
| 8 | 1,275 | <LD | <LD | <LD | >LD | 0,32 | <LD |
| 8' | 2,925 | <LD | <LD | <LD | 25,400 | 0,17 | <LD |
| V8 | 1,425 | <LD | <LD | <LD | >LD | 0,31 | 104,5 |
| 9 | 3,575 | <LD | <LD | <LD | >LD | 0,31 | <LD |
| 9' | 0,725 | <LD | <LD | <LD | 35,100 | 0,28 | <LD |
| V9 | 0,225 | <LD | <LD | <LD | 14,300 | <LD | 79,5 |
| 10 | 0,475 | <LD | <LD | <LD | 14,550 | 0,20 | <LD |
| 10' | 1,525 | <LD | <LD | <LD | 20,250 | 0,22 | <LD |
| V10 | 2,875 | <LD | <LD | <LD | 20,020 | 0,33 | <LD |
| 11 | 0,975 | NA | NA | NA | 6,125 | <LD | <LD |
| 11' | 0,975 | NA | NA | NA | 7,760 | <LD | <LD |
| V11 | 0,525 | NA | NA | NA | 6,360 | <LD | - |
| 12 | 0,675 | NA | NA | NA | 8,420 | <LD | <LD |
| 12 | 0,725 | NA | NA | NA | 6,345 | <LD | <LD |
| V12 | 1,425 | NA | NA | NA | 9,865 | 4,50 | <LD |
| 13 | <LD | NA | NA | NA | 3,685 | 4,00 | 0,0 |
| 13 | <LD | NA | NA | NA | 5,060 | <LD | 0,0 |
| V13 | 1,225 | NA | NA | NA | 16,430 | 4,50 | 69,0 |
| 14 | 0,000 | NA | NA | NA | 3,140 | 7,50 | 0,0 |
| 14' | <LD | NA | NA | NA | 6,960 | 6,00 | <LD |
| V14 | <LD | NA | NA | NA | 8,385 | 0,00 | <LD |
| 15 | <LD | NA | NA | NA | 5,390 | 0,00 | 0,0 |
| 15' | <LD | NA | NA | NA | 4,620 | 0,00 | 0,0 |
| V15 | <LD | NA | NA | NA | 0,000 | 0,00 | <LD |
| 16 | 0,000 | NA | NA | NA | 4,165 | <LD | 0,0 |
| 16' | 0,000 | NA | NA | NA | 3,335 | <LD | 0,0 |
| V16 | <LD | NA | NA | NA | 6,750 | 7,50 | <LD |

Caption:

- = lost sample

N.A = not analyzed

<LD = below the detection limit of the method

>LD = above the detection limit of the method

This confirmed the presence of these elements in the fish analyzed. However, as the values were very close to the method's detection limit, the samples were diluted to 25.0 mL instead of 50.0 mL as an adaptation of the method described by Anvisa and Ial (2005). Table 10 shows the method's detection limit values specified by the Instrumental Characterization Laboratory II (IQ/UERJ).

Table 10 - Method detection limit values (mg/L)

| Trace element | Concentration (mg/L) |
|---|---|
| Copper | 0,02 |
| Zinc | 0,015 |
| Lead | 0,08 |
| Aluminum | 1,2 |
| Nickel | 0,2 |
| Chrome | 0,16 |
| Cadmium | 0,03 |

### 4.3 Individual information

A total of 44 individuals were analyzed, excluding the individuals analyzed in the pilot test. The smallest specimen measured 7.5 cm, had a total weight of 5.45 g and a viscera weight of 0.41 g, while the largest measured 9.6 cm, had a total weight of 10.67 g and a viscera weight of 1.13 g. The average size was 8.9± 0.5 cm, the average total weight was 8.36± 1.15 g and the viscera weighed an average of 0.76± 0.18 g. The weight of the viscera contributed a minimum of 6.12%, a maximum of 12.47% and an average of 9.06 ±1.69% of the total weight of each individual. The weight and measurement values are listed in Table 11.

Table 11 - Individuals' weight and measurements

| Individual | Size (cm) | Total weight (g) | Viscera weight (g) | % Weight viscera |
|---|---|---|---|---|
| Minimum | 7,5 | 5,45 | 0,41 | 6,12 |
| Maximum | 9,6 | 10,67 | 1,13 | 12,47 |
| Average | 8,9 | 8,36 | 0,76 | 9,06 |
| Desv Pad | 0,5 | 1,15 | 0,18 | 1,69 |

### 4.4 Values of copper (mg/kg), zinc (mg/kg), lead (mg/kg) and aluminum (mg/kg) found in muscle and viscera

Copper in muscle showed an average concentration of 0.5± 0.66 mg/Kg and 1.3± 10.13 mg/Kg in viscera. Zinc values in muscle and viscera were 5.2± 3.69 mg/Kg, 25.6± 48.16 mg/Kg, respectively. The concentration of lead was 2.48± 3.09 mg/Kg (muscle) and 25.6± 48.16 mg/Kg (viscera), while the concentration of aluminum ranged from 1.68± 3.67 mg/Kg in muscle and 28.72± 26.99 mg/Kg in viscera.

Among the individuals analyzed, twenty-five samples (56.8%) showed values above the limit established for human consumption by Brazilian legislation (table 3) for Pb.

Despite the 239 industries and domestic sewage that is dumped into the Jacarepaguà lagoon, it has low concentrations of zinc, lead and copper in the water: 0.009± 0.003 mg/L; 0.0006± 0.0004 mg/L and 0.0009± 0.0004 mg/L, respectively (FERNANDES, 1994), and can be classified as a class 1 body of water according to CONAMA resolution 357/2005. In 1997, Fernandes studied the concentrations of iron, manganese, zinc, lead, nickel, copper and chromium in the sediment of the Jacarepaguà lagoon at six sampling points. At the sampling point closest to the site where this study was carried out, the concentrations of zinc, lead and copper were 507 mg/kg, 76 mg/kg and 82 mg/kg, respectively. The study observed that zinc and lead did not vary in their distribution throughout the lagoon. Thus, there were no significant changes in their distribution in the environment. As a result, Fernandes (1997) concludes that these are the most dangerous elements in terms of biota contamination in the Jacarepaguà lagoon.

Adeyeye and colleagues (1996) researched zinc, lead, manganese, iron, sodium and potassium in the head, eyes, gills, liver, intestine, reproductive organs, bladder, scales and muscle tissue of African catfish, carp and tilapia, as well as monitoring the water and soil. The results were that zinc, manganese, iron, sodium and potassium showed an uneven distribution between the various parts of the samples. Zinc in muscle tissue showed less variability. The tilapia samples had values of 3.93± 0.03 mg/kg in muscle and 1.35± 0.01 mg/kg in viscera. These values were found in an environment where the concentration of zinc in water was 0.03± 0.01 mg/kg and 0.15 ± 0.0 mg/Kg in sediment. Thus, these values were classified by the author as low zinc values in fish.

In this study, zinc had an average value of 5.2 ± 3.69mg/kg in muscle and 25.6 ± 48.16mg/kg in viscera. However, the concentration of zinc and lead in the water of Jacarepaguâ lagoon is lower than in the previous study. This is explained by the availability of zinc to the biota, as described by Fernandes (1997).

With regard to lead, the concentration found in viscera was 0.79± 0.02 mg/kg and the muscle, water and sediment samples showed undetermined values (ADEYEYE et al, 1996). In this study, the average lead concentrations found in Sardinella brasiliensis were 2.48 ± 3.09 mg/kg in muscle and 5.83± 11.72 mg/kg in viscera, which was also shown by Fernandes (1997) in fish that occupied the same trophic level as sardines in Lagoa da Tijuca. The literature also mentions that the larger the fish, the lower the total lead concentration due to higher metabolic rates (BAHNASAWY et al, 2009). Table 12 shows the mean values and standard deviation of trace element concentrations in muscle, viscera and total concentration.

Table 12 - Mean values and standard deviation of trace element concentrations in muscle, viscera and total concentration

| Part analyzed | Concentration (mg/Kg) | | | |
|---|---|---|---|---|
| | Copper | Zinc | Lead | Aluminum |
| Muscle | 0,5 ± 0,66 | 5,2 ±3,69 | 2.48± 3.09 | 1,74 ±3,6 |
| Viscera | 1,3± 10,13 | 25,6 ±48,16 | 5.83± 11.72 | 27,28 ±27,19 |
| Total | 0,79 ± 1,04 | 7,29 ± 5,72 | 3.15± 3.11 | 3.86± 4.35 |

The values found, when compared with other studies (Table 2), were close to those evaluated by Türkmen et al. (2009) for the species *B. belone*, which occupies the same trophic level as *Sardinella*

*Brasilienses.*

In their study, Türkmen et al. (2009) determined the concentrations of cadmium, cobalt, chromium, copper, iron, manganese, nickel, lead and zinc in the muscle and viscera of twelve fish species from the Aegean and Mediterranean Seas. In this study they concluded that the concentrations of metals in muscle are generally lower than in viscera and that although the fish do not present a risk for human consumption, the concentrations found indirectly indicate that the region was impacted.

Abdallah & Abdallah (2008) studied the concentrations of cadmium, copper, cobalt, zinc, manganese and iron in biota and sediment collected in El-Mex Bay and Eastern Harbour in the Mediterranean Sea. In this work, the authors use the table below (table 13) to compare their results.

Table 13 - Comparison of heavy metal concentrations in fish with other studies

| Area of study | Concentration (µg/g) | | | | | |
|---|---|---|---|---|---|---|
| | Cd | Cu | Co | Zn | Mn | Fe |
| Black Sea Coast[a] | <0,02 | 1,01 -4,54 | <0,05 -0,40 | 25,7 -44,2 | 0,69 -3,56 | 33 - 60 |
| Masan Bay - Korea[b] | 0,01 | 0,18 - 0,25 | - | 6,33 - 12,9 | - | - |
| Mediterranean Sea[0] | 0,37 -0,79 | 2,19 -4,40 | | 16,5 -37,4 | | 19,6 - 78,4 |
| Iskenderun Bay[d] | 0,95 | 1,57 | 1,42 | 4,36 | 1,71 | 10,2 |
| Aegean Sea[e] | <0,01 -0,04 | <0,1 | | <0,5 -7,2 | | |
| Eastern Harbour[f] | 1,7 | 3,07 | 2,4 | 16,29 | 0,82 | 24,1 |
| El-Mex Bay[g] | 2,32 | 2,56 | 3,47 | 30,97 | 3,87 | 41,8 |

Caption:

= -trace element not analyzed

[a]Topcuoglu et al apud Abdallah & Abdallah, 2008.

[b]Kwon and Lee apud Abdallah & Abdallah, 2008.

[c]Canli and Atli apud Abdallah & Abdallah, 2008.

[d]Türkmen et al apud Abdallah & Abdallah, 2008.

[e]Dalman et al apud Abdallah & Abdallah, 2008.

[f, g]Abdallah & Abdallah, 2008.

Source: ABDALLAH & ABDALLAH, 2008.

In all the studies cited, the highest concentration of zinc found was 44.2 mg/kg, while in the present study the highest concentration was 222.72 mg/kg. The average zinc level was 7.29 ± 5.72 mg/kg, which is close to the average value found in Massan Bay, Korea, and the average copper level in this study is similar to the average found on the Black Sea coast, both of which are reported to be impacted by heavy metals.

Canli and Atli (2003) studied the concentration of trace elements (cadmium, chromium, copper, iron, lead and zinc) in six species of fish, including *Sardina pilchardus*. This species consumes phytoplankton and zooplankton and is popularly known as the European sardine. The aim of this study was to evaluate the concentrations of these metals and analyze the correlation between size, weight and metal concentration in the tissues analyzed. The values found by the authors showed that the correlation between metal concentration, size and weight of the fish was negative, with a few exceptions. They also reported that trace element concentrations were high. In *Sardina pilchardus*, the concentration of copper in muscle was 4.17± 0.58 mg/Kg, lead 5.57± 1.03 mg/Kg and zinc 34.58± 8.64 mg/Kg. In viscera, the concentrations were 29.26 ± 13.19 mg/Kg copper, 39.43 ± 17.91 mg/Kg lead and 73.22 ± 19.84 mg/Kg zinc. This study also concluded that there was an increase in the concentration of lead as the size of the individual decreased. However, for copper and zinc there was no relationship with size, as these elements are regulated and maintained by metabolism up to certain concentrations.

Although the average concentrations of *Sardinella brasiliensis* were not close to those found in *Sardina pilchardus*, when looking at the maximum and minimum values, the values were close. This was due to the high difference in the standard deviation of the samples in this study, although there was a range of results in common.

The average lifespan of metallic compounds in the body is variable because it tends to be prolonged due to their protein affinity. The average lifespan of lead in human organisms can be over 20 years (BENITE et al, 2007).

Among the metals analyzed, zinc has the highest concentration allowed for ingestion (50 mg/kg) under Brazilian legislation. This may be due to the fact that this element does not contain any extremely toxic compounds (COTTON et al, 1999).

There were no samples above the limit established by Brazilian legislation for zinc (50 mg/kg) or copper (30 mg/kg). A total of 25 individuals (56.8%) were above the legal limit for lead (2 mg/Kg).

As for aluminum, 24 individuals were found to have high values when compared to the other 20 individuals who had low aluminum concentrations, which represents an increase in the concentration of aluminum in this species.

According to current legislation, the fish under analysis was unfit for human consumption due to high levels of lead. The absorption of lead in the gastrointestinal tract is around 15 to 20% for adults and up to 50% for children. Lead is absorbed through a calcium-binding protein, where a competitive effect with calcium can occur (TAVARES, 1992).

The values found should serve as a warning of future aluminum contamination of the *Sardinella brasiliensis* population that inhabits the outlet of the Sernambetiba canal.

It should be noted that fish that live in part or all of their lives in the ocean have low concentrations of metals (BAHNASAWY et al, 2009), which was not observed in all the specimens.

**4.5 Feature Grouping**

4.5.1 Size, weight and viscera weight

According to figure 11, the size of each individual was grouped with its weight, since the relationship between size and weight influences the typical shape of each species. This relationship was also found by Gobbi (2007) when he studied the fish species popularly known as Surubim. The grouping of the size and total weight of the individuals also indicated that all the mass acquired by the fish was converted into an increase in size, while also maintaining the typical shape of the species and showing that the individuals collected were young. This can be seen in Table 14, where the average size, 8.9 ± 0.5 cm, and the average weight, 8.36 ± 1.15 g, showed a low standard deviation.

Since each specimen has a viscera weight according to its needs, there is a variation in the absorption of trace elements from the diet, where despite the individuals having a low standard deviation in relation to weight (1.15 g) and size (0.5 cm) there is a standard deviation of 7.91mg/Kg in the average concentration of copper; 48.8 mg/Kg for the average zinc concentration; 11.72 mg/Kg for lead and 27.68 mg/Kg for aluminum.

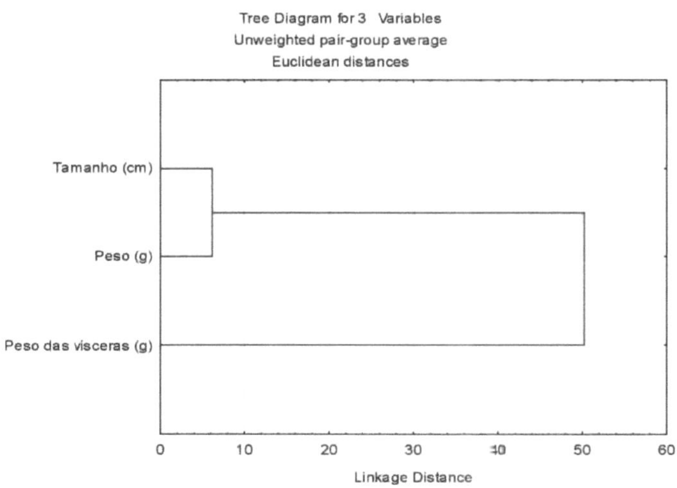


Figure 11 - Grouping for the characteristics of size, weight and viscera weight

4.5.2 Size, weight, viscera weight and trace element concentration

Terra (2008) and Bordajandi et al (2003) also observed higher concentrations of metals in viscera than in muscle. Thus, muscle appears to be the last preference for bioaccumulating metals (BAHNASAWY et al, 2009).

Porto and Ethur (2009) studied the concentration of trace elements in three families of fish from different trophic levels in the Butui-Icamaquã River Basin and found no correlation between the mass and/or length of the fish and the values of metals in them, except for nickel in fish from the Butui River.

However, the study by BAHNASAWY et al (2009) found a negative correlation between size and metal levels in most of the species studied (Mullet, Mugil Cephalus and Liza Ramada). The authors showed that there was a positive correlation between fish size and the concentration of zinc and lead in viscera.

In this study, the weight of the viscera correlated with the concentration of copper, aluminum and lead in the muscle, which means that the entry of the contaminant is related to the digestive tract and, consequently, the diet. This is further reinforced by the fact that Sardinella brasilliensis is not a benthic animal, so it does not come into direct contact with the contaminant present in the sediment. There was no correlation between the concentration in muscle and the concentration in viscera, as seen in figures 12 to 14. This result was also observed by Gobbi (2007).

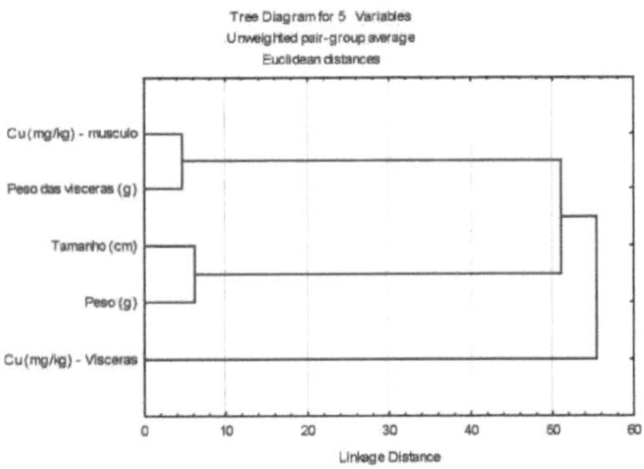


Figura 12 - Grouping for the characteristics of size, weight, viscera weight and copper concentrations in viscera and muscle

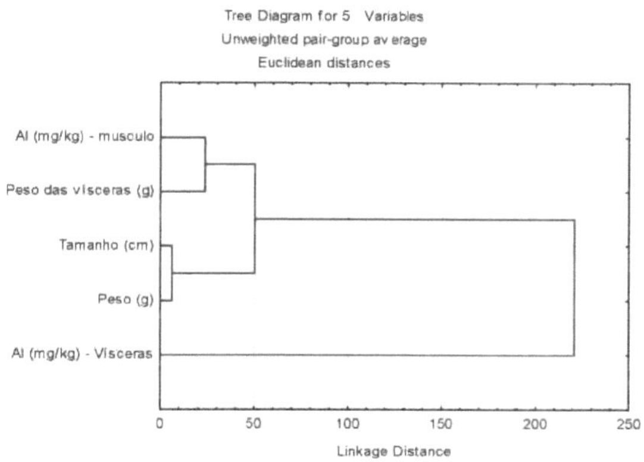


Figura 13 - Grouping for the characteristics of size, weight, viscera weight and aluminum concentrations in viscera and muscle

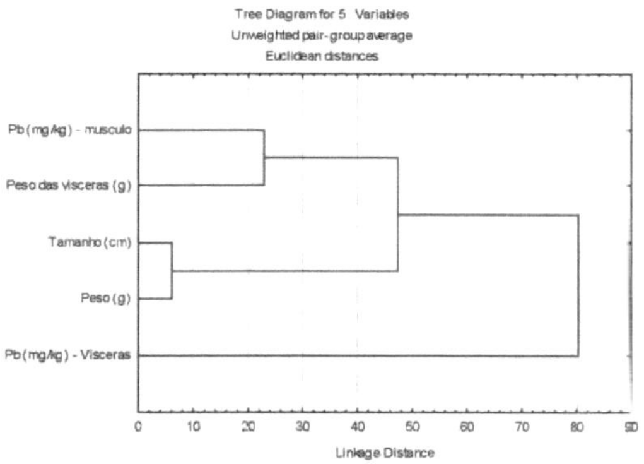


Figura 14 - Grouping for the characteristics of size, weight, viscera weight and lead concentrations in viscera and muscle

However, for the grouping of the same zinc characteristics, the parameters did not show the same pattern of variation (figure 15). The concentration of zinc in muscle formed a separate group from the weight of viscera and the concentrations in viscera. This is due to the ability to form complexes with ammonia within the digestive system and the good solubility of zinc compounds in water (COTTON et al, 1999).

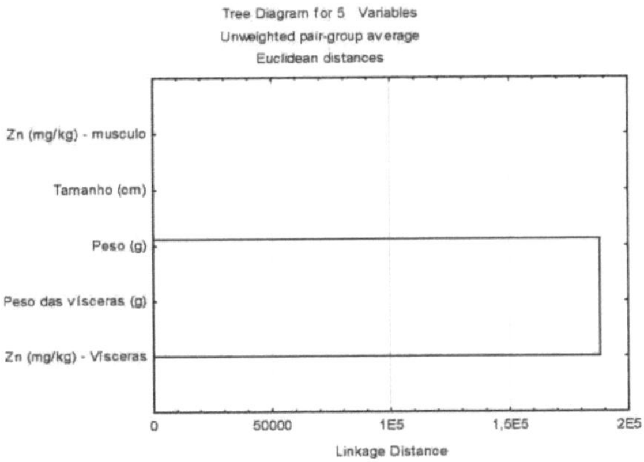


Figura 15 - Grouping for the characteristics of size, weight, viscera weight and zinc concentrations in viscera and muscle

4.5.3 Concentration of each trace element in total and in viscera

When the variation of metals in muscle and viscera is analyzed, there is a clear separation of these parameters (Figure 16). This showed that there is no correlation between the variation in muscle concentration and the variation in viscera concentration. The difference in concentration between the sites may be a consequence of the ability of some metalloproteins to induce metal shielding (BAHNASAWY et al, 2009; CANLI & ATLI, 2003).

Thus, as metalloproteins are present in organs with a detoxification function, they are found in the digestive and/or excretory systems. Because they are capable of retaining and mitigating the potential damage caused by xenobiotic agents, they end up accumulating the metals (xenobiotics) and allowing the smallest amount to be transferred to other tissues, such as muscle tissue.

We can also see that the order of the groups was different between the muscle concentration group and the viscera concentration group, which means that depending on the place of accumulation, the metals correlate differently. However, in both groups, zinc was the most distant from the others.

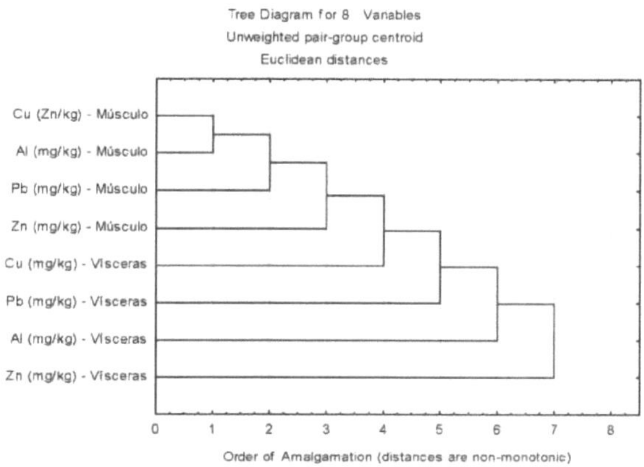


Figure 16 - Grouping for viscera concentration and muscle concentration characteristics

4.5.4 Concentration of each trace element in muscle and viscera

In the muscle, copper and aluminum form a separate group from lead and zinc (Figure 17), while in the viscera a single group is formed for copper, lead and aluminum, which differs from zinc (Figure 18). This shows that the trace elements do not tend to accumulate in the viscera and muscle in the same way.

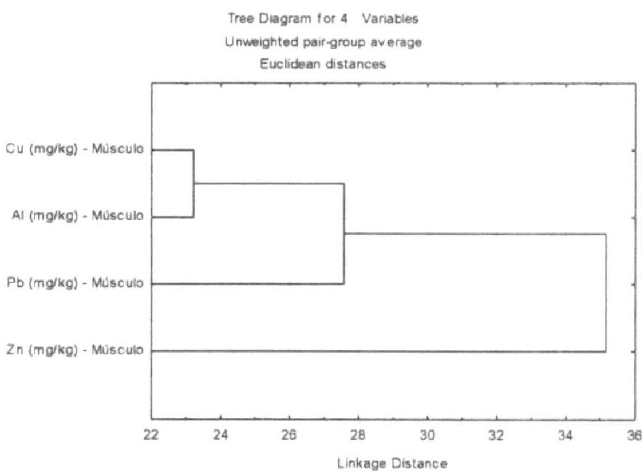


Figure 17 - Grouping of muscle concentrations

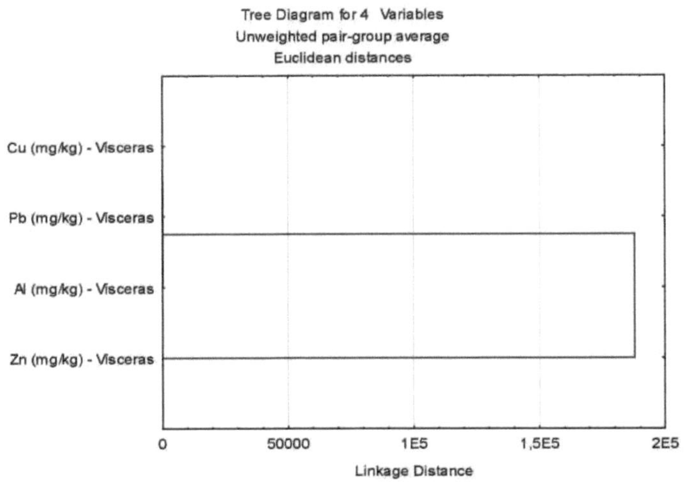


Figure 18 - Grouping of viscera concentrations

# 5 CONCLUSIONS

Among the individuals analyzed, 56.8% of the samples showed values above the limit established for human consumption by Brazilian legislation for lead, although its impact on human health depends on various factors such as the participation of fish and other sources of these elements in the population's diet.

The concentrations of trace elements in *Sardinella brasilliensis* are similar to those found in Massan Bay, Korea and the Black Sea (all areas impacted by heavy metals). Thus, the concentrations found in this study suggest that the region is impacted.

The individuals showed a variation in the absorption of trace elements from the diet. Although the individuals had a low standard deviation in terms of weight and size, the percentage ratio between the weight of the individual and the weight of the viscera varied more, as did the concentration of heavy metals, demonstrating that individual characteristics had a significant influence on the accumulation of metals in Sardinella brasiliensis.

Fish growth and its relationship with metal concentration in the aquatic environment should be monitored to increase understanding of the effects of metals on fish development and population dynamics.

# 6 SUGGESTIONS FOR FUTURE WORK

Below are some suggestions for future work that would complement the work done in this dissertation. They are:

- Seeking a spatial-temporal characterization of the water quality of the Tijuca-Jacarepaguâ-Marapendi lagoon system and its current trophic state index.

- Carry out a survey of the industries in the region and the quality of their effluents in order to study the environmental impact caused by them on the ecosystem.

- To carry out an assessment of toxic metals in sediment, water and various trophic levels of biota, in order to create an up-to-date database of the region studied, which can also be used as a parameter for other studies of trace metal contamination.

- Research into the compartments of this aquatic ecosystem in order to understand how the metal cycle works in Jacarepaguâ Lagoon.

# 7 BIBLIOGRAPHY


ACQUA FORUM. **National Program for Safety, Health and the Working Environment in Fishing and Professional Diving Activities.** 2007 Available at: <http://www.fundacentro.sc.gov.br/acquaforum>. Accessed on: June 13, 2010.

ADAMS, S. M (ed). Biological indicators of aquatic ecosystem stress: Introduction and overview. **Biological Indicators of Aquatic Ecosystem Stress.** American Fisheries Society, Maryland, p. 1-11, 2002.

ADEYEYE, E.I.; AKINYUGHA, N.J.; FESOBI, M.E.; TENABE, V.O. Determination of some metals in Clarias gariepinus (Cuvier and Valenciennes), Cyprino Carpio and Oreochromis niloticus fishes in a Polyculture Freshwater Pond and their Environments. **Aquacult.**, v.147, p. 205-214, 1996.

ALLEN, H. E.; HANSEN, D. J. The importance of chemical speciation to water quality criteria. **Water Environ. Res.,** v.68, p.42, 1996.

ALONSO, E.; SANTOS, A.; CALLEJÓN, M.; JIMÉNEZ, J.C. Speciation as a screening tool for the determination of heavy metal surface water pollution in the Guadiamar river Basin. **Chemosphere,** v.56, p.561, 2004.

ANVISA - National Health Surveillance Agency. **Ordinance No. 685 of August 27, 1998.**

ANVISA - National Health Surveillance Agency. **Specific legislation for the food sector - inorganic contaminants (heavy metals) in fish and fishery products,** 2003. Available at: http://www.anvisa.gov.br.

ANVISA - National Health Surveillance Agency. **Decree No. 55.871, of March 26, 1965**. Available at: http://anvisa.gov.br/legis/resol/2000/23 00.htm.

ANVISA - National Health Surveillance Agency; IAL - Adolfo Lutz Institute. Minerals and inorganic contaminants. In: ZENEBON, O. et al. (Coord.). **Métodos fisico-quimicos para anàlise de alimentos**. Sao Paulo: ANVISA- IAL, 4.ed., chap.23, p.739-758, 2005.

ARAUCO, L.R.R.; CRUZ, C.; NETO, J.G.M. Effect of the presence of sediment on the acute toxicity of copper sulfate and trichlorfon to three species of daphnia. **Pesticides: ecotoxicol. e meio ambiente,** v. 15, p.55-64, 2005.

ASUQUO, F. E.; EWA-OBOHO, I.; ASUQUO, E. F; UDO, P. J. Fish species used as biomarker for heavy metal and hydrocarbon contamination for Cross river, Nigeria. **The Environmentalist**, v.2, p. 29-37, 2004.

ASUQUO, F. E.; OGRI, O. R.; BASSEY, E. S. Distribution of heavy metals and total hydrocarbons in coastal waters and sediments of cross River State, South Eastern Nigeria. **International Journal of Tropical Environment**, v.2, p. 229-242, 1999.

AWWA - American Water Works Association. Alum sludge in the aquatic environment. **Research Foundation and American Water Works Association**, 224 p., 1991.

BAHNASAWY, M.; KHIDR, A. A.; DHEINA, N. Seasonal Variations of Heavy Metals Concentrations in Mullet, Mugil Cephalus and Liza Ramada (Mugilidae) from Lake Manzala, Egypt. **Journal of Applied Sciences Research,** v.5, n. 7, p.845-852, 2009.

BAIRD, C. **Quimica ambiental.** 2. ed. Porto Alegre: Bookman, 2002.

BARBOSA, R. M.; POVINELLI, J.; ROCHA, O.; ESPiNDOLA, E. L. G. **Anais do XXVII Congresso Interamericano de Ingenieria Sanitaria y Ambiental**, Porto Alegre, 2000.

BARCAROLLI, I.F.; MARTINEZ, C.B.R. Effects of aluminum in acidic water on hematological and physiological parameters of the neotropical fish Leporinus macrocephalus (Anostomidae), **Bull. Environ. Contam. Toxicol.,** v.72, p. 639-646, 2004.

BARROSO, L.V.; MEDINA, R.S.; MOREIRA-TURCQ, P.F.; BERNARDES, M.C. **Série Meio Ambiente em Debate: A pesca nas lagoas costeiras fluminenses, Intituto Brasileiro do Meio Ambiente e dos Recursos Naturais Renovàveis (IBAMA)**, Diretoria de Gestao Estratégica. Brasilia: Ed. IBAMA, 50 p., 2000.

BERTOLLETI, E. Toxicity and concentration of toxic agents in industrial effluents. **Ciência e cultura**, Sao Paulo, v.14, n.3, p.271-277, 1990.

BORDAJANI, L. R.; GÓMEZ, G.; FERNANDEZ, M. A.; ABAD, E.; RIVERA, J.; GONZALEZ, M. J. Study on PCBs, PCDD/Fs, organochlorine pesticides, heavy metals and arsenic content in freshwater fish species from River Turia (Spain). **Chemosphere**, v.53, p.163-171, 2003.

BRAGA, B.; HESPANHOL, I.; CONEJO, J.G.L.; MIERZWA, J.C.; BARROS, M.T.L.; SPENCER, M.; PORTO, M.; NUCCI, N.; JULIANO, N.; EIGER, S. **Introdução à engenharia ambiental: o desafio do desenvolvimento sustentâvel**. 2ªEd. Sao Paulo, 315p., 2005.

BRAZIL. Laws, Decrees. **Decree no. 55.871 of March 26, 1965 of the Ministry of Health.** Establishes maximum limits for inorganic contaminants in food. Official Journal of the Union. Brasilia, DF, May 29, 1965.

BRESOLA, R. C. Toxicity assessment of water sources in areas degraded by mining using the bioindicator *Scenedesmus subspicatus* and implementation of toxicity methodology with the popular fish *Danio rerio* "zebrafish". **Master's dissertation,** 84p, 2007.

BRUN, A. E.; DIETZ BRUNE, K.-J. ; DIETZ. A comparative analysis of element composition of roots and leaves of barley seedlings grown in the presence of toxic cadmium, molybdenum, nickel, and zinc concentrations. **Journal of Plant Nutrition**, v. 18, n.4, p. 853 - 868, 1995.

CALDERÓN-SALINAS, J.V., VALDEZ, A.B., MAZÛNIGA, C., ALBORES, M. A. Lead exposure in a population of Mexican children. **Hum. Exp.Toxicol.**, v.15, p.305-311, 1996.

CALLEGARI-JACQUES SM. **Biostatistics: principles and applications.** Porto Alegre: Artmed, 2003.

CAMPOS, M. L. A. M.; BENDO, A.; VIEL, F. C. Low-cost methods for reagent purification and contamination control for the determination of trace metals in natural waters. **Quim. Nova**, v. 25, n. 5, p. 808-813, 2002.

CANLI, M.; ATLI, G. The relationships between heavy metal (Cd, Cr, cu, Fe, lead, Zn) levels and the size of six Mediterranean fish species. **Environmental Pollution**, v. 121, p. 129-136, 2003.

CASARINI, D.C.P.; DIAS, C.L.; LEMOS, M.M.G. Report on the establishment of guideline values for soil and groundwater in the state of Sao Paulo. **CETESB - IV Series (Environmental Reports Series)**. Sao Paulo, p.73, 2001.

CHAGAS, G.C. Evaluation of the bioindicator potential of trichodactylus fluviatilis (latreille, 1828) (crustacea: decapoda: trichodactylidae) in the corumbatai river basin. **Master's Degree** in Biological Sciences (Zoology), 2008.

COELHO, E.P.M.; VIEIRA, M.F.P.; MOURA, M.F.V.; MEDEIROS, R.L.S.; PANERÒ, F.S.; SILVA, H.E.B.; FARIAS, R.F. Multivariate analysis in food chemistry: determination of heavy metals in mullet (*mugil brasiliensis*) in the state of Rio Grande do Norte. **Quimica no Brasil,** v. 1, n. 2, 2007.

CONAMA. **RESOLUTION 020-1986.** Ministry of the Environment, National Environment Council. 1986. Available at: http://www.mma.gov.br/port/conama/res/res86/res2086.html.

CONAMA. **RESOLUTION 357-2005.** Ministry of the Environment, National Environment Council. 2005. Available at: http://www.mma.gov.br/port/conama/res/res05/res35705.pdf.

CORREIA, TIAGO G. Influence of aluminum and acidic pH on the reproductive physiology of continental teleost fish. **Master's Degree in Biosciences**, 204 p., 2008.

CORRER, C.J.; MAZZOCHIN, L.F.; LOBO, I.; OTAVIO, J.G.A.S.; GUEDES, C.L.B. "In situ" application of surfactants in petroleum-contaminated soil and treatment of wastewater with an oxidizing agent. **Proceedings of the 4th PDPETRO**, 2007.

COSTA, E. C. T. A. Environmental diagnosis of the waters of the Jundiai- Potengi estuary through the determination of water quality and toxicity indices. **Master's dissertation in Chemistry,** 135p., 2008.

COSTA, J. R. Distribution of metals in marine fish along the southeast coast of Brazil. 2007. 43p. **Monograph in Biological Sciences - Emphasis in Environmental Sciences.**

COTTON, F. A.; WILKINSON, G.; MURILLO, C. A.; BOCHMANN, M. **Advanced Inorganic Chemistry**, 6th ed., Willey-Interscience Publication, 1999.

CYPRIANO, E.F.; ARAÙJO, C.L.; LACERDA, L.D.; LOUREIRO, D.D. Spatial distribution of mercury in the sediments of the Baixada de Jacarepaguà lagoon complex (RJ). 2008. **Proceedings of the**

**III Brazilian Congress of Oceanography - CBO and I Ibero-American Congress of Oceanography - I CIAO.**

DEMUELENAERE, R.G.A. Characterization of Heavy Metal Transport Properties in Residual Soils from Rio de Janeiro. 2004. 120p. **Master's Degree in Civil Engineering.**

DOMINGOS, F.X.V. Biomarkers of environmental contamination in fish and oysters from three Brazilian estuaries and kinetics of soluble petroleum derivatives in fish. 2006. 130p. **PhD in Biological Sciences with a concentration in Cellular and Molecular Biology.**

U.S. EPA - **Environmental Protection Agency**, 2008. Available at: <http://water.epa.gov/scitech/swguidance/waterquality/standards/criteria/aqlife/dissolved/dorefs.cfm>. Accessed on February 10, 2010.

ESTEVES, F. DE A. **Fundamentals of Limnology.** Rio de Janeiro. Interciência, 2ed., 1998.

FAIRMAN, B.; SANZ-MEDEL, A. Field Sampling Technique for the "Fast Reactive" Aluminum Fraction in Waters Using a Flow Injection Mini-column System with Inductively Coupled Plasma Atomic Emission Spectrometric and Inductively Coupled Plasma Mass Spectrometric Detection. **Journal of Analytical Atomic Spectrometry,** v.355, p. 757-762, 1995.

FERNANDES, H.M. Heavy metal distribution in sediments and ecological risk assessment: the role of diagenetic processes in reducing metal toxicity in bottom sediments. **Environmental Pollution,** v. 97, n. 3, p. 317-325, 1997.

FERNANDES, H.M.F.; BIDONE, E.D.; VEIGA, L.H.S.; PATCHINEELAM, S.R. Heavy metal: pollution assessment in the coastal Lagoons of Jacarepaguâ, Rio de Janeiro, Brazil. **Environmental Pollution,** v.85, p.259-264, 1994.

FERNANDES, V.O. Successional studies of the periphytic community in leaves of Typha dominguensis Pers, related to limnological variables in Jacarepaguâ Lagoon, RJ. 1993. 131p. **Master's dissertation.**

Food and Agriculture Organizations of the United Nations (FAO). World Review of Fisheries and Aquaculture, Fisheries Resources: Trends in Production, Utilization and trade, 2003. **Technical report.** Available at: <https://www.fao.org.br>. Accessed on: March 20, 2010.

Food and Agriculture Organizations of the United Nations (FAO/WHO). Summary of evaluations performed by the joint FAO/WHO expert committee on food additives (JECFA 1956-2003), 2004. **Technical report.** Available at: <https://www.fao.org.br>. Accessed on: March 20, 2010.

GENSEMER, R.W.; PLAYLE, R.C. The Bioavailability and Toxicity of Aluminum in Aquatic Environments. **Critical Reviews in Environmental Science and Technology**, v.29, n.4, p. 315 - 450, 1999.

GOBBI, J.M. Study on the presence of metals in different tissues of surubim fish (Pseudoplatystoma

Coruscans) caught in the São Francisco River (MG). 2007. 60p. **Master's dissertation.**

GONÇALVES, G. O. Distribution of Mercury in the Fluminense Lagoon System, Niterói, RJ. 1999. **Master's thesis in Geochemistry.**

GUNDERSEN, P.; STEINNES, E. Influence of pH and TOC concentrations on Cu, Zn,Cd and Al speciation in rivers. **Water Res.,** v.37, p.307, 2003.

HARRIS, D.C. **Quantitative chemical analysis.** 7th ed. United States of America: W.H.Freeman and Company, 661p., 2007.

HAS-SCHON, E.; BOGUT, I.; STRELEC, I. Heavy metal profile in five fish species included in human diet, domiciled in the end flow of river Neretva (Croatia). **Archives of Environmental Contamination and Toxicology**, v.50, p. 545-551, 2006.

HEATH, A.G. **Water pollution and fish phisiology**. Virginia: Lewis Publishers, 359 p., 1995.

HENRIQUES, G.S.; HIRATA, M.H.; COZZOLINO, S.M.F. Recent aspects of zinc absorption and bioavailability and correlations with physiology of the testicular Angiotensin-Converting Enzyme. **Rev. Nutr**., v. 16, n. 3, p. 333-345, 2003.

HENRY, F.; AMARA, R.; COURCOT, L.; LACOUTURE, D.; BERTHO, M. L. Heavy metals in four fish species from the French coast of the Eastern English Channel and Southern Bight of the North Sea. **Environmental International**, v.30, p.675-683, 2004.

HOLT, M.S. Sources of chemical contaminants and routes into the freshwater environment. **Food Chem. Toxicol**., v.38, p.21-27, 2000.

IBAMA - Brazilian Institute for the Environment and Renewable Natural Resources. **Technical Report: Environmental Aspects and Fishing Activities in Fluminense Coastal Lagoons,** 2000.

JAC-70-0002 RE: Environmental impact study for the Jacarepaguà macrobasin environmental recovery project. **Diagnosis of the physical environment**, 1998.

JESUS, S.C. Zinc diffusion in a compacted layer of residual gneiss soil. 2004. 90p. **PhD thesis in Civil Engineering.**

LACERDA, L. D.; FERNANDEZ, M. A.; CALAZANS, C. F.; TANIZAKI, K. F. Bioavailability of heavy metals in sediments of two coastal lagoons in Rio de Janeiro, Brazil. **Hydrobiologia,** v.228, p.65-70, 1992.

LACERDA, L.D.; GONÇALVES, G.O. Mercury Distribution and Speciation in Waters of the Coastal Lagoons of Rio de Janeiro, SE Brazil. **Marine chemistry**, v.76, n.1-2, 47-58, 2001.

LARINI, L. **Toxicology.** Sao Paulo: Manole, 315p, 1993.

LEMES, M. J. L., FIGUEIREDO FILHO, P. M.; PIRES, M. A. F. Influence of the mineralogy of sediments from the Mogi-guaçu and Pardo river basins on the chemical composition of public water supplies. **Quim. Nova**, v. 26, n. 1, p.13-20, 2003.

LINHARES, S.; GEWANDSAJDER, F. **Biologia hoje**. Sao Paulo. Ed. Atica. 424p., 2003.

LU, Y.; ALLEN, H. E. Partitioning of copper onto suspended particulate matter in river waters. **Sci. Tot.Environ**. v.277, p.119-132, 2001.

MAGALHAES, V.F.; SOARES, R.M.; AZEVEDO, S.M.F.O. Microcystin contamination in fish from the Jacarepaguà Lagoon (Rio de Janeiro, Brazil): ecological implication and human health risk. **Toxicon**, v. 39, p. 1077-1085, 2001.

MAVROPOULOS, E. Hydroxyapatite as a lead remover. 1999. 126p. **Master's dissertation in Public Health.**

MELETTI, P.C.; MARTINE, O.R.; REIS, C.B. Evaluation of environmental degradation in the Mogi-Guaçu River Basin by means of sediment toxicity tests and histopathological analysis in fish. **Limnologia Fluvial: um estudo no Rio Mogi-Guaçu**, Chap 9, p.65, 2003.

MENEZES, A.C. Toxicity of copper on tambaqui colossoma *macropomum* (cuvier, 1818) at ph 4 and ph 8. 2005. 81p. **Master's dissertation.**

MERCOSUR - Southern Common Market, accessed at:http://www.mercosul.gov.br/, Resolution No. 102/94, 1994.

MEYLAN, S.; BEHRA, R.; SIGG, L. Accumulation of copper and zinc in periphyton in response to dynamic variations of metal speciation in freshwater. **Environ. Sci. Technol**. v.37, p.5204- 5212, 2003.

MUEHE, D. ; VALENTINE, E. The Brazilian coastline and its compartmentalization. Ed.Guerra. **Geomorphology of Brazil.** Rio de Janeiro, 1998.

NASCENTES, R. Study of the mobility of heavy metals in a compacted residual soil. 2006. 153p. **Gold thesis.**

NOGUEIRA, R.F.P.; JARDIM, W.F. Heterogeneous photocatalysis and its environmental application. **Chem. Nova**, v.21, n.1, p. 69-72, 1998.

ODUM, E. P. **Ecology.** Rio de Janeiro. Ed. Guanabara Koogan S.A., 434p., 1983.

OLIVEIRA, M.S.R.O. Estudo da especiaçâo de metais traços em sedimentos de um tanque de cultivo de camarao marinho do município de Santa Rita-Pb. 2006. 69p. **Master's dissertation.**

WHO. **Trace elements in human nutrition and health.** Sao Paulo. Ed. ROCA, 297p., 1998.

PEREIRA, A.A.F.; TENUTA-FILHO, A. Evaluation of consumption conditions for Sardines *Sardinella brasiliensis*. **Ciênc. Tecnol. Aliment**. v.25, n.4, p.720725, 2005.

PINO, G. A. H. Biosorption of heavy metals using green coconut shell powder (*Cocos nucifera).* 2005. 113p. **Master's Degree in Materials Science and Metallurgy.**

PORTO, L.C.S.; ETHUR, E.M. Trace elements in water and fish viscera from the Butui-Icamaqua Hydrographic Basin, Rio Grande do Sul, Brazil. **Ciência Rural**, v.39, n.9, p.2512-2518, 2009.

PRADIT, S.; WATTAYAKORN, G.; ANGSUPANICH, S.; BAEYENS, W.; LEERMAKERS, M. Distribution of Trace Elements in Sediments and Bicta of Songkhla Lake, Southern Thailand. **Water Air Soil Pollut,** v.206, p.155-174, 2010.

PTASHYNSKI, M.D.; PEDLAR, R.M.; EVANS, R.E.; BARON, C.L.; KLAVERKAMP, J.F. Toxicology of dietary nickel in lake whitefish (Coregonus clupeaformis). **Aquatic Toxicology,** v.58, n.3-4, p.229-247, 2002.

RASHED, M. N. Monitoring of environmental heavy metals in fish from Nasser Lake. **Environment International,** v.27, p.27-33, 2001.

REZENDE, P. S. DE. Evaluation of the distribution and mobility of trace elements in sediments from the São Francisco River basin. 2009. **Master's Degree in Chemistry - Analytical Chemistry.**

ROCHA, J.C.; ROSA, A. H.; CARDOSO, A.A. **Introdução à quimica ambiental.** Ed.: Bookman, 2004.

RUIZ-FERNANDEZ, A. C., et al. The Loading History of Trace Metals and Nutrients in Altata-Ensenada del Pabellon, Lagoon Complex, Northwestern Mexico. **Journal of Environmental Radioactivity**, v.69, n.1-2, 2003.

RUSSEL, John Blair. **General chemistry.** Translation and technical review by Màrcia Guekezian et al. 2. ed. Sao Paulo: Makron Books, 1994.

SALDANHA, T.; BENASSIB, M.T.; BRAGAGNOLO, N. Fatty acid contents evolution and cholesterol oxides formation in Brazilian sardines (Sardinella brasiliensis) as a result of frozen storage followed by grilling, **ScienceDirect**, n.41, p. 1301-1309, 2008.

SALGADO, P.E.T. Toxicology of metals. In: Seizi, O. **Fundamentos de toxicologia**. Sao Paulo: Ateneu, p. 150-152, 1996.

SANTANA, E.Q. Determination of macroelements, trace elements and metallic contaminants in propolis by flame and graphite furnace atomic absorption spectrophotometry. 2003. **PhD thesis in Pharmaceutical Sciences.**

SANTOS, D.C.M. Acute zinc toxicity in lambaris Astyanax Bimaculatus (Linnaeus, 1758). 2009. **Master's thesis in Animal Biology**.

SCHNEIDER, F. & P.R. SCHWINGEL. Preliminary study of the trophic ecology of Sardinella brasiliensis off the southeast coast of Brazil. **Notas técnicos. facimar**, v.3, p.67-72, 1999.

SEMADS - State Secretariat for the Environment and Sustainable Development. **River basins and rivers in Rio de Janeiro: Informative Synthesis by Environmental Macroregion.** Semads Water Plan Project / Brazil-Germany Technical Cooperation Group, 2001.

SIGG, L. Regulation of trace elements in lakes: The role of sedimentation. In: Buffle, J.; De Vitre, R. R. (Ed), **Chemical and Biological Regulation of Aquatic Systems**. Ed. Boca Raton, p177-197, 1994.

SORENSEN, E.M. **Metal Poisoning in Fish.** Ed.Boca Raton, 374 p., 1991.

TAO, S.; LONG, A.; XU, F.; DAWSON, R.W. Copper Speciation in the Gill Microenvironment of Carp (Cyprinus carpio) at Various Levels of pH. **Ecotoxicology and Environmental Safety.** v.52, p.221-226, 2002.

TARLEY, C.R.T.; COLTRO, W.K.T., MATSUSHITA, M.; SOUZA, N.E.S. Characteristic Levels of Some Heavy Metals from Brazilian Canned Sardines (Sardinella brasiliensis). **Journal of Food Composition and Analysis,** v.14, p. 611-617, 2001.

TAVARES, T.M.; CARVALHO, F.M. Evaluation of exposure of human populations to heavy metals in the environment: an example from the recôncavo baiano. **Chem. Nova,** n.15, p.147, 1992.

TERRA, B.F.; ARAÙJO, F.G.; CALZA, C.F.; LOPES, R. T.; TEIXEIRA, T. P. Heavy metal in tissues of three fish species from different trophic levels in a tropical brazilian river. **Water Air Soil Pollut,** v.187, p. 275-284, 2008.

TONIETTO, A.E. Chemical specification of copper and zinc in the waters of the irai reservoir: effects of organic matter and interaction with microalgae. 2006. 164p. **Master's dissertation.**

TRIEBSKORN, R.; TELCEAN, I.; CASPER, H.; FARKAS, A.; SANDU, C.; STAN, G.; COLĂRESCU, O.; DORI, T.; KOHLER, H. Monitoring pollution in River Mureș, Romania, part II: Metal accumulation and histopathology in fish. **Environ Monit Assess,** v.141, p. 177-188, 2008.

TSUTIYA, M.T. Heavy metals: The main limiting factor for the agricultural use of biosolids from sewage treatment plants. In: **Proceedings of the Brazilian Congress of Sanitary and Environmental Engineering,** p.753-761, 1999.

TURAN, C.; DURAL, M.; OKSUZ, A.; OZTÜRK, B. Levels of heavy metals in some commercial fish species captured from the black sea and Mediterranean coast of Turkey. **Bull Environ Contam Toxicol,** v.82, p. 601-604, 2009.

TÜRKMEN, A.; TEPE, Y.; TÜRKMEN, M.; MUTLU, E. Heavy Metal Contaminants in Tissues of the Garfish, Belone belone L., 1761, and the Bluefish, Pomatomus saltatrix L., 1766, from Turkey Waters. **Bull Environ Contam Toxicol**, v. 82, p. 70-74, 2009.

TÜRKMEN, M.; CIMINLI, C. Determination of metals in fish and mussel species by inductively coupled plasma-atomic emission spectrometry. **Food Chem,** v.103, p. 670-675, 2007.

TWISS, M. R.; ERRÉCALDE, O.; FORTIN, C.; CAMPBELL, P. G. C.; JUMARIE, C.; DENIZEAU,F.; BERKELAAR, E.; HALE, B.; VAN REES, K. Coupling the use of computer chemical speciation models and culture techniques in laboratory investigations of trace metal toxicity. **Chem. Spec.**

**Bioavail.** n.13, p.9, 2001.

VALITUTTO, R.S.; SELLA, S.M.; SILVA-FILHO, E.V.; PEREIRA, R.G.; MIEKELEY, N. Accumulation of metals in macrophytes from water reservoirs of a power supply plant, Rio de Janeiro State, Brazil. **Water Air Soil Pollut,** v.178, p.89-102, 2006.

VAN HULLEBUSCH, E. D. , ZANDVOORT, M. H.; LENS, P. N. L. Metal immobilization by biofilms: Mechanisms and analytical tools. **Re/Views in Environmental Science & Bio/Technology,** v.2, p. 9-33, 2003.

VANZ, A.; MIRLEAN, N.; BAISCH, P. Assessment of air pollution by particulate lead: A geochemical approach. **Quimica Nova**. v. 26, n.1,p.25-29. 2003.

VILA NOVA, C.M.V.M.; GODOY, H.T.; ALDRIGUE, M.L. Composiçao quimica da tilâpia **e** do pargo, **Ciência e Tecnologia de Alimentos**, v. 25, n.3, p.430436, 2005.

WHO. Environmental Health. **Criteria 101: Methylmercury.** Geneva: World Health Organization, 1990.

WHO Environmental Health. **Food Additives Series: Safety evaluation of certain food additives and contaminants,** 2000.

WIENER, J.G.; SPRY, D.J. Toxicological significance of mercury in fresh water fish. In BEYER, W.N. et al. **Environment contaminants in wildlife: interpreting tissue concentrations.** Boca Raton: Lewis Publishers, Chap.13, p.297-339, 1996.

WREN, C.D.; STEPHENSON, G.L. The effect of acidification on the accumulation and toxicity of metals to freshwater invertebrates. **Environmental Pollution.** v.71, n.2-4, p.205-241, 1991.

ZAGATTO, P.A. Ecotoxicology. In: ZAGATTO, P.A.; BERTOLETTI, E. (Eds.). **Aquatic Ecotoxicology - Principles and Applications.** Ed. Sao Paulo, SP, Brazil. Editora Rima, 2006.

ZHOU, J.L.; SALVADOR, S.M.; LIU, Y.P.; SEQUERIA, M. Heavy metals in the tissues of dolphins (Delphinus delphis) stranded on the Portuguese coast. **Sci Total Environ,** v.273, p.61-76, 2001.

ZSOLNAY, A. Dissolved organic matter: artifacts, definitions and functions. **Geoderma** , v.113, p.187 - 209, 2003.

MIX
Papier aus verantwortungsvollen Quellen
Paper from responsible sources
FSC® C105338

Printed by Books on Demand GmbH, Norderstedt / Germany